6th Asia Conference on Mechanical Engineering and Aerospace Engineering (MEAE 2020)

MATEC Web of Conferences Volume 316 (2020)

Chengdu, China
14 – 16 June 2020

Editor:

C.H. Su

ISBN: 978-1-7138-1444-3

Printed from e-media with permission by:

Curran Associates, Inc.
57 Morehouse Lane
Red Hook, NY 12571

CURRAN ASSOCIATES INC. proceedings.com

Some format issues inherent in the e-media version may also appear in this print version.

For additional information, please contact EDP Sciences – Web of Conferences at the address below.

EDP Sciences – Web of Conferences
17, Avenue du Hoggar
Parc d'Activité de Courtabœuf
BP 112
F-91944 Les Ulis Cedex A
France

Phone: +33 (0) 1 69 18 75 75
Fax: +33 (0) 1 69 28 84 91

contact-edps@webofconferences.org

Additional copies of this publication are available from:

Curran Associates, Inc.
57 Morehouse Lane
Red Hook, NY 12571 USA
Phone: 845-758-0400
Fax: 845-758-2633
Email: curran@proceedings.com
Web: www.proceedings.com

TABLE OF CONTENTS

PREFACE

This volume contains papers presented at 2020 6th Asia Conference on Mechanical Engineering and Aerospace Engineering (MEAE 2020), which was held during June 14-16, 2020 in Chengdu, China.

MEAE 2020 provides a scientific platform for both local and international scientists, engineers and technologists who work in all aspects of Mechanical Engineering and Aerospace Engineering. The goal of MEAE is to establish an international forum for engineers and scientists to present their ideas and experiences in the fields of Mechanical Engineering and Aerospace Engineering like we always do.

The volume includes 18 selected papers which were submitted to the conference from universities, research institutes and industries. Each contributed paper has been peer-reviewed by reviewers who were collected organizing and technical committee members as well as other experts in the field from different countries. The proceedings tend to present to the readers the newest researches results and findings in the field of Mechanical Engineering and Aerospace Engineering.

The chairpersons of different sessions played important role in conducting the proceedings of the session in a timely and efficient manner, and on behalf of the conference committee, we express sincere appreciation for their involvement. The reviewers of manuscripts have also been very helpful in efficiently reviewing the manuscripts, providing valuable comments well within the time allotted to them. We express our sincere and grateful thanks to all the sponsors and participants for their valuable contributions and support to MEAE 2020.

Finally, we wish you have a stimulating and fruitful time at the conference.

MEAE-2020 Conference Committee

Assessment of Material Dislocation Damage by Nonlinear Ultrasound

Zaifu Zhan[1], Shen Wang[1,a], Songling Huang[1], Yang Zheng[2], Fuping Wang[1] and Wei Zhao[1]

[1]State Key Laboratory of Power System, Department of Electrical Engineering, Tsinghua University, 100084 Beijing, China
[2]Inspection and Research Institute, China Special Equipment, 100084 Beijing, China

Abstract. Under harsh environment or during service, the mechanical properties of materials or structure will deteriorate. Most of the simulations exhibit the phenomenon of nonlinearity by introducing the actual small defects, without considering dislocation. In this manuscript, subroutines are written to change the mechanical constitutive behaviour of materials. When the mechanical constitutive behaviour of the material is not linear any more, it is found that the propagation of ultrasonic wave in the material will show more obvious nonlinear phenomenon. Furthermore, the nonlinear detection coefficient is used to characterize the increase of harmonic components. This work provides a new idea for nonlinear ultrasonic testing.

1 Introduction

Since structure under periodic load or in harsh environment, no matter what materials, will undergo material properties degradation. Continuous degradation leads to fail of structures, sometimes with serious consequences. The evaluation of structural changes of materials and constructions and monitoring their ultimate strength and endurance in operation is required for many industrial structural integrity problems. Especially the evaluation of accumulated damage or degradation in material properties at the early stage of fracture is important in refinery plants, nuclear power plants, or aircraft parts [1-2].

The early degradation of the mechanical properties takes up most of the life of the structures. Especially, the research shows that the degradation process includes the accumulation of the dislocation inside the materials, formation of the persistent slip band and finally the appearance of the cracks. This process accounts for 80%~90% [3-4] of the life of the whole structure when the metal material structure is bearing periodic load. Therefore, testing of the early properties of materials is particularly important.

The most powerful nondestructive way of evaluating material degradation is the ultrasonic method since the characteristics of ultrasonic wave propagation are directly related to the properties of the material. Traditional ultrasonic nondestructive evaluation (NDE) is based on linear theory and normally relies on measuring some particular parameter (sound velocity, attenuation, transmission and reflection coefficients) of the propagating signal to determine the elastic properties of a material or to detect defects [5]. The presence of defects changes the phase and/or amplitude of the output signal, but the frequency of the input and output signals is the same.

However, the conventional ultrasonic technique is sensitive to gross defects or opens cracks, where there is an effective barrier to transmission, whereas it is less sensitive to evenly distributed microcracks or degradation[6]. An alternative technique to overcome this limitation is nonlinear ultrasonics[7-12]. The principal difference between linear and non-linear ultrasound NDE is that in the latter the existence and characteristics of defects are often related to an acoustic signal whose frequency differs from that of the input signal. This is related to the radiation and propagation of finite amplitude (especially high power) ultrasound and its interaction with discontinuities, such as cracks, interfaces and voids. Since material failure or degradation is usually preceded by some kind of nonlinear mechanical behaviour before significant plastic deformation or material damage occurs, considerable attention has recently been focused on the application of nonlinear ultrasonics[13].

When the material is undamaged, it obeys the linear constitutive relation. The wave propagation in the material is linear. But when the material is damaged, the mechanical constitutive behaviours is no longer linear any more. The constitutive relation of damaged materials is extremely complex[12-15]. This led to distortion of wave form and appearance of harmonics in the process of wave propagation.

Many researchers have done the research in nonlinearities of the structure response after ultrasonic excitation. Buck[16] studied this theoretically and experimentally, and developed a pinned dislocation monopoles model of ultrasonic and dislocation effects. He pointed out that the generation of ultrasonic second harmonic depends on the displacement movement of a single dislocation between the pinning points which are

[a] Corresponding author: wangshen@mail.tsinghua.edu.cn

under stress. The theoretical formulas among the amplitude of the second harmonic, the stress and the length of the dislocation line are derived. Cantrell[17] et al proposed a new model of ultrasonic to interact with dislocation dipoles and dipole arrays for exciting second harmonics. They believed that when the dislocation is dense, the amplitude of the second harmonic of the ultrasound is not only related to the length and stress of the dislocation lines, but also largely influenced by the arrangement of dislocations in the structure.

However, few researchers simulate the interact between microscopic dislocations and ultrasound. Most scholars ignore this earlier dislocation stage and study crack directly, like Ref [2, 3, 6, 8, 18, 19, 21]. This manuscript simulates the damage before cracks appear and conclude that ultrasound can already reflect early dislocations of material damage, which will be an important guarantee for NDE in military and aerospace.

The remainder of this manuscript is organized as follows: Section 2 introduces not only the basic theoretical background of nonlinear ultrasound and higher harmonics generation in theory but also constitutive behaviour which are written for simulations. In Section 3, a finite element model(FEM) of the propagation of nonlinear longitudinal waves is described in detail. FEM simulation results are presented and discussed in Section 4, and conclusions are drawn in Section 5.

2 Basic Theoretical Background

In this section, the nonlinear wave equation is derived from two methods. It can be seen from the derivation that if the material no longer satisfies the linear Hooke's law, the wave equation is also nonlinear, which leads to the generation of nonlinear phenomena. Therefore, the constitutive relationship is changed in simulation for damaged materials. The changed constitutive relation also will be given.

2.1 Nonlinear wave equation

This is the general expression for the free energy of a deformed isotropic body [14]:

$$F=F_0+\frac{1}{2}\lambda\varepsilon_{ii}^2+\mu\varepsilon_{ik}^2 \quad (1)$$

where, the coefficients λ and μ are called Lamé coefficients, F_0 is the initial free energy value, ε_{ik} is the strain tensor whose linear expression is [14]:

$$\varepsilon_{ik}=\frac{1}{2}\left(\frac{\partial u_i}{\partial x_k}+\frac{\partial u_k}{\partial x_i}\right) \quad (2)$$

where, u_i is particle displacement in i direction and x_1, x_2 and x_3 are x, y, z directions respectively in Cartesian coordinate system.

When we consider nonlinearity, the strain tensor should be expanded [14]:

$$\varepsilon_{ik}=\frac{1}{2}\left(\frac{\partial u_i}{\partial x_k}+\frac{\partial u_k}{\partial x_i}+\frac{\partial u_l}{\partial x_k}\frac{\partial u_l}{\partial x_i}\right) \quad (3)$$

If only one-dimensional situation is considered, the linear wave equation:

$$\frac{\partial^2 u}{\partial t^2}-c\frac{\partial^2 u}{\partial x^2}=0 \quad (4)$$

should be amended to:

$$\frac{\partial^2 u}{\partial t^2}-c(1-\beta\frac{\partial u}{\partial x}+\epsilon\,(\frac{\partial u}{\partial x})^2)\frac{\partial^2 u}{\partial x^2}=0 \quad (5)$$

where, c is wave celocity, β and ϵ are the quadratic and cubic nonlinear coefficients, respectively.

There is another simpler theoretical explanation that when the material is undamaged, the relationship between stress and strain is linear:

$$\sigma=E\varepsilon \quad (6)$$

However, when the material is damaged, the relationship between stress and strain is no longer linear:

$$\sigma=E\varepsilon(1+\beta\varepsilon+\ldots) \quad (7)$$

where, σ is stress. Substituting this relation into one-directional wave equation:

$$\rho\frac{\partial^2 u}{\partial t^2}=\frac{\partial\sigma}{\partial x} \quad (8)$$

Then we can obtain nonlinear wave equation:

$$\rho\frac{\partial^2 u}{\partial t^2}=A\frac{\partial u}{\partial x}+B\frac{\partial u}{\partial x}\frac{\partial^2 u}{\partial x^2} \quad (9)$$

Cantrell [4, 20] et al. also established similar nonlinear ultrasonic wave equation:

$$\rho\frac{\partial^2 u}{\partial t^2}=k_2\frac{\partial u}{\partial x}+(3k_2+k_3)\frac{\partial u}{\partial x}\frac{\partial^2 u}{\partial x^2} \quad (10)$$

2.2 Mechanical constitutive behaviour

If we only consider the x direction, the stress-strain relationship obeys eq. (6). Its components can be expressed in the form of:

$$\sigma_{ij}=\lambda\delta_{ij}\varepsilon_{kk}+2\mu\varepsilon_{ij} \quad (11)$$

where δ_{ij} is the Kronecker Symbol, $\delta_{ij}=1$ if $i=j$ and $\delta_{ij}=0$ if $i\neq j$.

For simplicity, we only consider the quadratic constitutive relation:

$$\sigma=E\varepsilon(1+\beta\varepsilon) \quad (12)$$

The code of subroutine VUMAT needs the form of component, which is

$$\sigma_{ij}=\left(\lambda\delta_{ij}\varepsilon_{kk}+2\mu\varepsilon_{ij}\right)\left(1+\beta\lambda\delta_{ij}\varepsilon_{kk}+2\beta\mu\varepsilon_{ij}\right) \quad (13)$$

Because the Explicit solver of Abaqus software will be used in the following simulation, recursive eq. (14) is required. Other details are described in the next section.

$$\Delta\sigma_{ij}=\left(\lambda\delta_{ij}\Delta\varepsilon_{kk}+2\mu\Delta\varepsilon_{ij}\right)\left(1+\beta\lambda\delta_{ij}\Delta\varepsilon_{kk}+2\beta\mu\Delta\varepsilon_{ij}\right) \quad (14)$$

https://doi.org/10.1051/matecconf/202031601001

In this manuscript, the following work will begin with the linear stress-strain relation of materials and change the stress-strain relation in materials by writing subroutines, so as to simulate the undamaged and damaged materials.

3 Numerical Model

Two-dimensional FE models were built with Abaqus and dynamic simulations were conducted with Explicit solver.

The Explicit solver is specialized in solving complex problems such as Explicit dynamic analysis within short time. It does not solve the coupled equation of system or generate the total stiffness matrix in each increment step, but recurs results of the previous step in a small increment step in time domain. It only calculates the total energy in the simulation to determine whether there is divergence in each increment step. So we need the recurrence eq. (14) of constitutive relation in the second section.

Abaqus software provides a user Subroutine interface-VUMAT for User to define the mechanical constitutive behaviour of a material. Therefore, the constitutive relationship of materials is changed by writing user subroutine VUMAT, in which β is set to 0, 0.0001, 0.0002, 0.0003, 0.0004 and 0.0005 respectively for simplicity. When β equals 0, it means the material is undamaged.

The 2D model is 40mm long and 20mm thickness. The excitation is applied in the left edge of model within the width of 6mm, and the excitation amplitude is 6μm, as shown in Fig. 1. The material properties are in table 1.

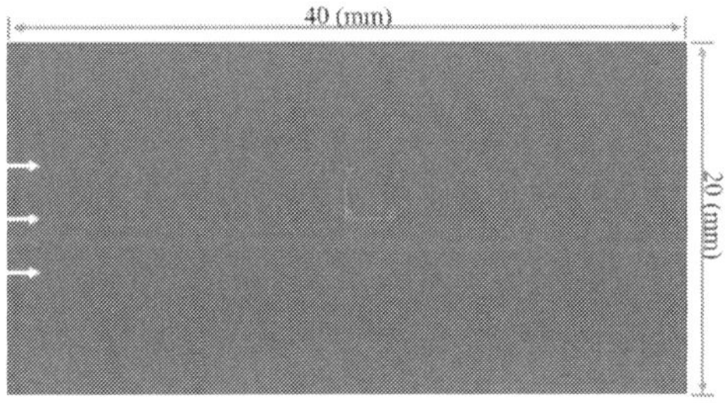

Figure 1. 2D model and excitation.

Table 1. Material properties.

Material type	Density	Young's modulus	Poisson's ratio
Steel	7800kg/m^3	2×10^{11} Pa	0.3

The excitation signal shown in Fig. 2 is a five cycle tone-burst. Window function is used to prevent spectrum leakage. The excitation signal central frequency is 3×10^6 Hz, and its expression is as follows:

$$y=\sin(2\pi ft)\,(0.08+0.46(1-\cos(2\pi ft/5))) \qquad (15)$$

The mesh size must be able to represent the physical characteristics of longitudinal wave propagation. So the smallest wavelength $\lambda_{\min}$ must be correctly sampled to describe the propagating mode in the frequency range for analysis. The spatial discretization of δx_1 and δx_2 of each element of the mesh satisfies the condition [21]:

$$\frac{\lambda_{\min}}{max(\delta x_1,\delta x_2)}>10 \qquad (16)$$

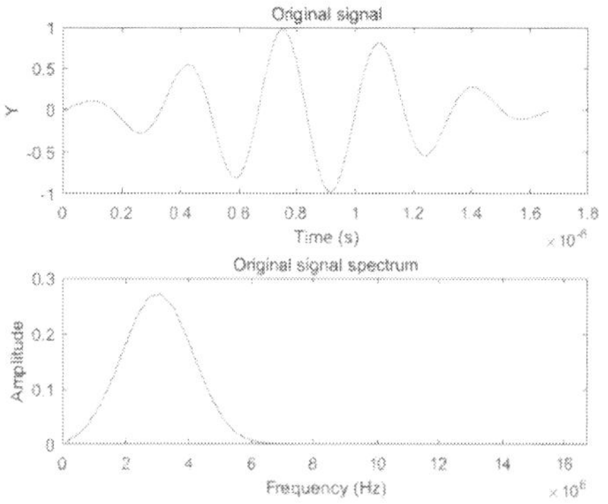

Figure 2. Excitation signal and spectrum.

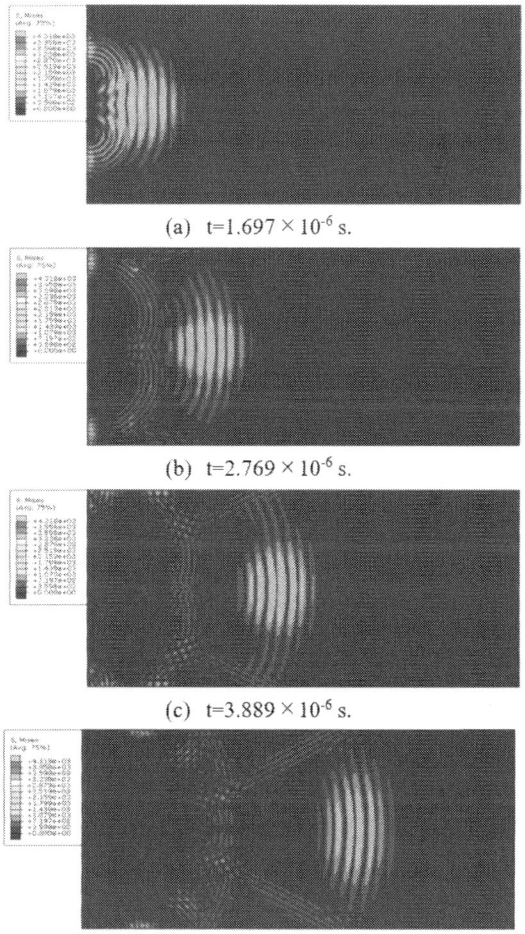

(a) t=1.697 × 10^{-6} s.

(b) t=2.769 × 10^{-6} s.

(c) t=3.889 × 10^{-6} s.

(d) t=5.2 × 10^{-6} s.

Figure 3. Stress nephogram at different times during ultrasonic propagation.

Due to the need to measure harmonics, the size of mesh elements is set to 5×10^{-5} m.

Abaqus software contains a wealth of mesh elements libraries. We choose CPE4R (i.e. linear reduction integral plane strain type) as the element type. It should be noted that the linear reduced integral refers to the linear interpolation used by the mesh elements in the calculation, which has no relations with the nonlinear source mentioned in this manuscript, but only for saving the calculation resources and meeting the algorithm of Abaqus/Explicit solver.

Another stability condition is that the time step must be chosen so that no wave can propagate across one mesh spacing in less than one-time step. Typically, the time discretization must satisfy the condition [21]:

$$\Delta t \leq 0.7\frac{min(\delta x_1,\delta x_2)}{c_l} \tag{17}$$

where c_l is the velocity of longitudinal wave. Therefore, in order to be precise, the step time is set as 1 ns, and the total simulation time is 8μs.

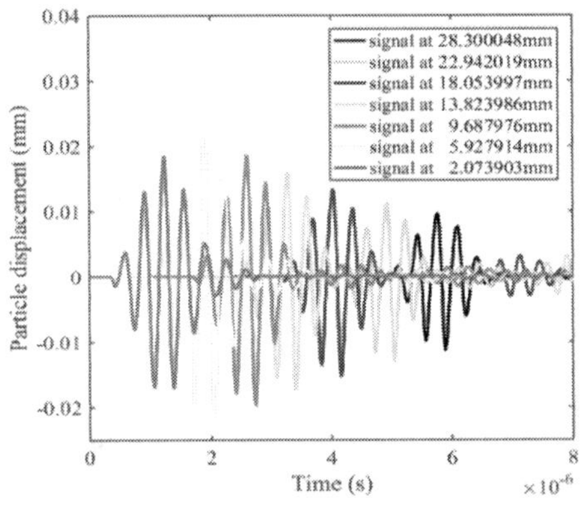

Figure 4. Waveform at different position in undamaged material model.

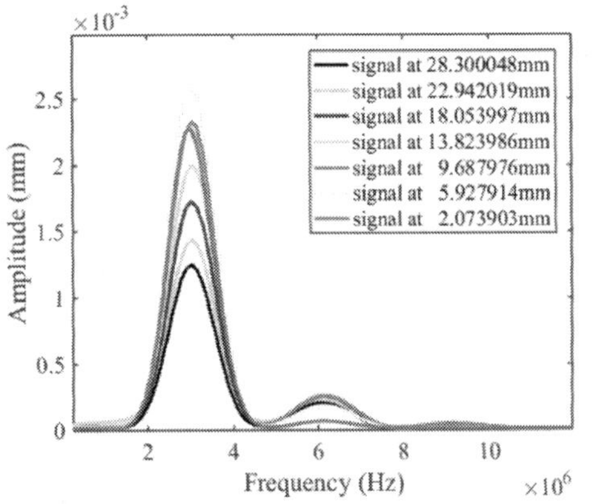

Figure 5. Spectrum of each curve in Fig. 4.

4 Simulation Result and Discussion

The wave propagation process is shown in Fig. 3 with stress nephogram for more intuitive. In the process of ultrasonic propagation, the displacement boundary condition excites not only the longitudinal wave but also the transverse wave. The propagation speed of transverse wave is obviously lower than longitudinal wave.

When the material is not damaged in simulations(i.e. β is set to zero or simulate without VUMAT), the constitutive relation in the material is linear, such as eq. (6) or eq. (11). Select several points on the propagation path of the material, and take the curve of the strain component ε_{11} (because the wave is a longitudinal wave) as shown in Fig. 4. In Fig. 5, spectrum analysis on each curve(signals at different distance) were performed for obtaining the fundamental, second and third harmonic components, as shown in the table 2. Combined with the stress nephogram in Fig. 3, we can clearly see that, in the process of wave propagation, the width of the wave is widened and the beam spreading phenomenon appears. Therefore, with the increase of the propagation distance, the energy is dispersed, and the particle displacement of the measurement point decreases with wave propagation.

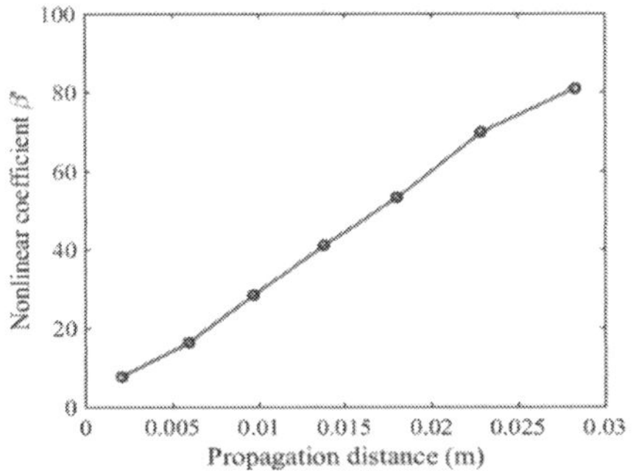

Figure 6. Variation of nonlinear detection coefficient β' with propagation distance in undamaged material.

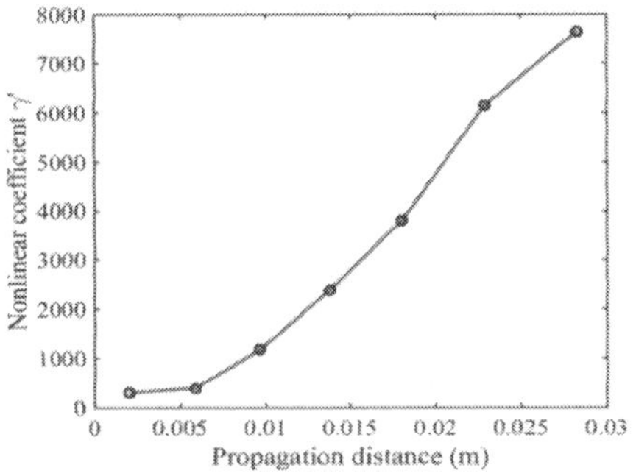

Figure 7. Variation of nonlinear detection coefficient γ' with propagation distance in undamaged material.

Table 2. Harmonic components at different positions.

Propagation distance (mm)	2.073903	5.927914	9.687976	13.823986	18.053997	22.942019	28.300048
Baseband amplitude (mm)	0.003605	0.004121	0.003691	0.003171	0.00274	0.002272	0.001977
Second harmonic amplitude (mm)	0.000099	0.000276	0.000386	0.000413	0.000399	0.00036	0.000316
Third harmonic amplitude (mm)	0.000014	0.000028	0.000059	0.000076	0.000078	0.000072	0.000059

In Fig. 4, we can find that there is small vibration after five periods of longitudinal wave propagation. The main reason is the extra transverse wave effect. Although there are additional shear waves, the shear wave frequency is the same as the original excitation frequency, so it will not affect the frequency analysis.

However, due to the phenomenon of beam spreading in the process of propagation, we cannot directly compare the harmonic magnitude. We need to further analyse by introducing β' and γ' of nonlinear ultrasonic detection coefficient, the expression is as follows [18]:

$$\beta' = \frac{A_2}{A_1^2} \quad (18)$$

$$\gamma' = \frac{A_3}{A_1^3} \quad (19)$$

where, A_i is the amplitude of the ith harmonic.

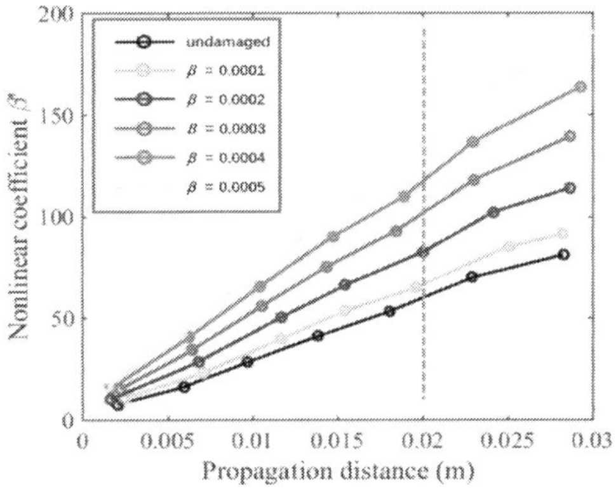

Figure 8. Variation of nonlinear detection coefficient β' with propagation distance in different damaged degree models.

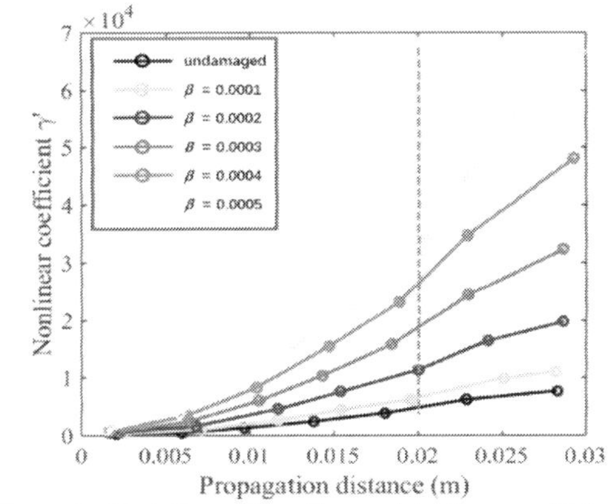

Figure 9. Variation of nonlinear detection coefficient γ' with propagation distance in different damaged degree models.

We analyze the spectrum of each curve data from table 2 and get the result as shown in Fig. 6 and Fig. 7. We can see that the nonlinear detection coefficient β' and γ' increases with the increase of propagation distance. This reflects the accumulation of harmonics in the process of propagation.

When we change the constitutive relation in materials, such as eq. (12). We have done five groups of simulation experiments. The nonlinear coefficient β is set to 0.0001, 0.0002, 0.0003, 0.0004 and 0.0005 respectively. The waveform and stress nephogram of damaged material is similar to that of undamaged material because β is relatively small.

The numerical simulation results are shown in Fig. 8 and Fig. 9. Different colours correspond to different β values. When β is determined (i.e. the damage degree of material is determined), we can see that the nonlinear detection coefficient of material increases with the increase of distance. The trend of increase is the same as when the material is undamaged. However, when the material is damaged, the nonlinear coefficient increases faster. This is a very obvious feature, the damage degree of two identical materials can be judged through it. We can get the reference value by measuring the undamaged material and then judge the damage degree of the material at a certain time by the reference value.

In Fig. 8 and Fig. 9, a virtual vertical line means a certain propagation distance. If the propagation distance is constant, the second detection coefficient β' is larger as the material damage is more serious.

In such a degree of mesh generation, the detection coefficient γ' obtained by measuring the third harmonic is highly sensitive to the damage of materials by comparing with β'. However, β' is a better quantity to show material damage by comparison of Fig. 8 and Fig. 9. Because there is a more obvious linear relationship between β' and propagation distance. This linear relationship provides a basis for quantitative measurement of material damage in practical engineering.

From the spectrum in Fig. 5, we can see, the amplitude of the third harmonic is extremely small. If the simulation mesh is not dense enough, it is easy to produce large errors. In the actual engineering, the measurement is difficult and the error is large because the amplitude of the third harmonic is too small. So the second harmonic is still the main method for applying nonlinear ultrasound to practical engineering.

Through the above simulations and analysis, we can see that in the uniformly damaged material, or before the crack occurs, the regional damage of materials in early stage can be detected.

5 Conclusions

Nonlinear respond of ultrasound is an important means to explore the damage degree of materials. As the mechanical properties of materials deteriorate, the constitutive relations of materials change. In this manuscript, the constitutive relation of simulation is changed by the method of writing a subroutine. It is found that early material degradation can be detected by ultrasound before the appearance of cracks. The generated harmonics are characterized by nonlinear detection coefficients β' and γ': the nonlinear detection coefficient increases as the distance of propagation increases, reflecting the accumulation effect of harmonics. In addition, the nonlinear coefficient increases with the increase of material damage degree when the propagation distance is fixed. This laid a foundation for damage determination of materials in the early dislocation stage.

References

1. Hong Gun Kim, Hee Jae Shin, Sun Ho Go, Min Sang Lee, Alexandre Tugirumubano, Yoo Bin Kim, Kyoung Soo Kim, and Lee Ku Kwac, International Journal of Mechanical Engineering and Robotics Research, **6**, 1 (2017)
2. C. R. Farrar and K. Worden, Philos Trans A Math Phys Eng Sci, **365**, 1851 (2007)
3. N. G. Meyendorf, H. R¨osner, V. Kramb, and S. Sathish, **40**, 8, (2002)
4. Cantrell and J. H, Proceedings of the Royal Society A: Mathematical, Physical and Engineering Sciences, **460**, 2043 (2004)
5. A. S. Birks, ASNT Handbook, 1991.
6. P. B. Nagy, Ultrasonics, **36**, 5 (1998)
7. A. Raghavan, Dissertation Abstracts International, **975**, 2 (2007)
8. I. Y. Solodov, N. Krohn, and G. Busse, Ultrasonics, **40**, 1 (2002)
9. I. K. Zarembo and V. A. Krasil'Nikov, Soviet Physics Uspekhi, **102**, 6 (1970)
10. K. Y. Jhang and K.-C. Kim, Ultrasonics, **37**, 1 (1999)
11. Y. Zheng, R. G. Maev, and I. Y. Solodov, Canadian Journal of Physics, **77**, 12 (1999)
12. D. Broda, W. Staszewski, A. Martowicz, T. Uhl, and V. Silberschmidt, Journal of Sound and Vibration, **333**, 4 (2014)
13. K. Y. Jhang, IEEE-UFFC, **47**, 3 (2000).
14. E. L.D. Landau, Oxford: Elsevier Ltd, **187** (1986)
15. K. E. A. V. D. Abeele and J. Carmeliet, Research in Nondestructive Evaluation, **12**, 31–42 (2000).
16. B. O., Review of progress in quantitative nondestructive evaluation, 1990.
17. J. H. Cantrell and W. T. Yost, Philosophical Magazine A, **69**, 2, (1994).
18. X. Wan, Q. Zhang, G. Xu, and P. Tse, Sensors, **14**, 8528–8546 (2014)
19. G. M. Fierro, F. Ciampa, D. Ginzburg, E. Onder, and M. Meo, Journal of Sound and Vibration, **343**, 121–130 (2015)
20. V. D. Abeele and K., The Journal of the Acoustical Society of America, **99**, 3 (1995).
21. J. Jingpin, M. Xiangji, H. Cunfu, and W. Bin, NDT & E International, **85**, 63–71 (2016).

Research on the Method of Multi-point Pressing Adjustment of Space Mechanism Driven by Measurement

Jinlong Zhao[1,2,a], Chunzhen Ren[1,2], Zhizhuo Cui[3] and Fuzhou Du[3]

[1]*Beijing Institute of Spacecraft Environment Engineering, Beijing, China*
[2]*Beijing Engineering Research Center of the Intelligent Assembly Technologyand Equipment for Aerospace Product, Beijing, China*
[3]*Beihang University, Beijing, China*

Abstract. In order to solve the problems of difficult assembly and adjustment and poor operation accessibility in the assembly process of multi-point pressure-type large-scale equipment on spacecraft, this paper presents a method of multi-point compact assembly of space mechanism driven by measurement. The method establishes the geometric constraint mathematical model of the equipment pressing mechanism, proposes the assembly coordinate system based on the mounting hole and the compensation method based on the rigid body transformation, and develops the relevant software according to the method. This method has been verified during the installation of the equipment and has achieved good results, which can be used as a reference for other digital assembly methods of aerospace equipment.

1 Introduction

In recent years, digital assembly technology has flourished, and because of its high precision, high efficiency and low risk, it is gradually replacing traditional assembly technology. At present, for the assembly features of different scale assemblies, the corresponding digital assembly technology can be used for efficient assembly operations, such as small-scale robot shaft hole assembly[1], large-scale wall panel flexible assembly[2], large parts digitization docking[3] and so on.

In the aerospace field, with the continuous development of manned space technology, the various types of equipment carried on the spacecraft have been greatly improved in terms of quantity, scale and functionality. Among them, there is a large-scale multi-point compression type large cabin. The equipment has the characteristics of large mass and size, multiple degrees of freedom, multiple joints, and uneven mass distribution. In order to ensure that the equipment does not damage during the launching process, the connection mode between the equipment and the spacecraft cabin is designed as a multi-point compression self-locking connection[4-5]. The self-locking compression mechanism locks the device in the ground assembly and launching state, and unlocks and releases the device after entering the rail, so that it enters the normal working state [6]. At present, the typical multi-point compression large-scale outboard equipment includes space manipulator [7-9], solar battery wing, etc.

Multi-point compression equipment assembly generally uses the position and attitude of the pressing mechanism that the gasket adjustment device cooperates with the cabin. The pressing mechanism in the assembly is distributed on different planes, the relative spatial position is complex, and the installation precision of each pressing mechanism is high. Secondly, due to the large scale of the assembly itself, the operator can't enter the inner side of the mating surface for operation, and can only be assembled by repeated trial and adjustment, so the assembly efficiency is low and some risks are always exist. In addition, there is a problem of poor accessibility during the installation and alignment of the equipment due to the relatively complicated layout position of the equipment pressing mechanism. Therefore, it is necessary to adopt a digital measurement-based scheme to guide the assembly of the pressing mechanism, and reduce the repeated cycle of the trial-and-adjustment during assembly, so assembly at one time can be achieved and the assembly efficiency and ensuring assembly quality are improved.

Aiming at the structural characteristics and assembly requirements of a certain type of multi-point pressure-type large-scale equipment on manned spacecraft, this paper proposes a measurement-driven multi-point compression fitting method to solve the problems of difficult installation and adjustment, poor accessibility and low efficiency exist in the loading process of this kind of equipment.

2 Multi-point compression assembly

2.1 The structure of the pressing mechanism

[a] Corresponding author: zhaojinlongcasc@163.com

Compression connection is a unique form of fastening connection, and a single compression mechanism is divided into an upper pressing seat and a lower pressing seat (also called release nut). Under normal circumstances, the upper pressing seat is self-contained, and the lower pressing seat is connected to the mounting surface of the device. When the equipment is installed, it is necessary to precisely fit the plurality of upper pressing seats on the equipment and the corresponding plurality of lower pressing seats on the mounting surface, and then perform locking by the locking device between the pressing seats. The basic structure of the pressing mechanism is shown in Figure 1 (the locking device is not shown).

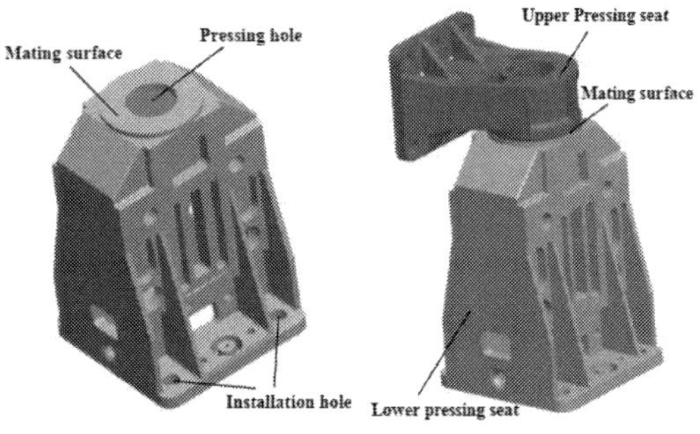

Figure 1. Pressing mechanism and connection diagram

2.2 Assembly requirements for the compression mechanism

The assembly form of the pressing mechanism is the cooperation of the two end faces of the upper and lower pressing seats, and the assembly requirements are mainly embodied in the gap of the mating surface and the relative position of the upper and lower circular holes. According to the assembly requirements, the corresponding mathematical relationship can be obtained, as shown in formulas (1) to (3), and the coordination relationship is shown in Figure 2.

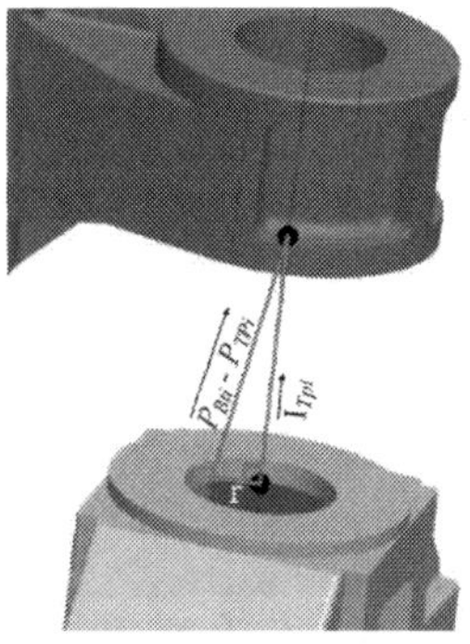

Figure 2. Schematic diagram of the cooperation mechanism of the pressing mechanism

$$\left|\overrightarrow{P_{Bti} - P_{Tpi}} \cdot \overrightarrow{I_{Tpi}}\right| < c \quad (1)$$

$$\sqrt{\overrightarrow{P_{Bti} - P_{Tpi}}^2 - \overrightarrow{P_{Bti} - P_{Tpi}} \cdot \overrightarrow{I_{Tpi}}^2} < r \quad (2)$$

$$\angle\left(\overrightarrow{I_{Tpi}} - \overrightarrow{I_{Bti}}\right) < \theta \quad (3)$$

In the formula, P_{Tpi} and P_{Bti} are the coordinates of the hole center of the pressing surface of upper and lower pressing seats on the mating surface, r is the coaxiality requirement of the fitting in the assembly, and is indirectly expressed by the deviation of the center position of the mating surface, and c is the clearance requirement of assembly fit, θ is the axial angle requirement for the clamping mechanism, $\overrightarrow{I_{Tpi}}$ and $\overrightarrow{I_{Bti}}$ are the normal vectors of the mating surfaces.

2.3 Working condition analysis of multi-point pressing assembly

Under normal circumstances, when multi-point pressure-type large equipment is assembled, the pressing mechanism is distributed in multi points and different planes in space, and the matching form is shown in Figure 3.

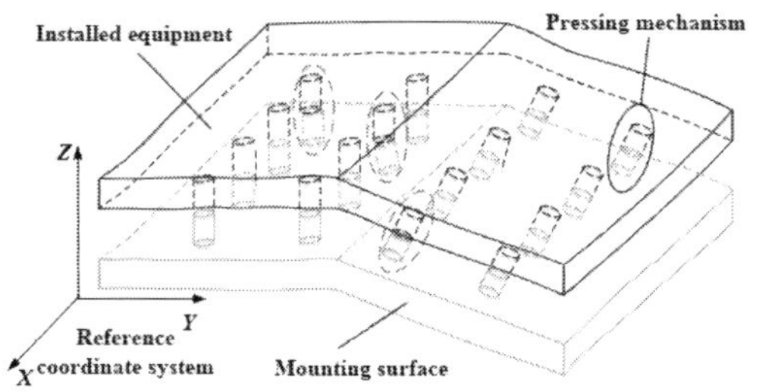

Figure 3. Pressing mechanism and connection diagram

When the lower pressing seat of the pressing mechanism is mounted on the equipment mounting surface, it is connected to the mounting surface through 4 bolt holes. Each mounting hole has a Φ 1mm translatable interval for adjusting the position, and the height and inclination can be adjusted properly by the gasket of 4 mounting holes, as shown in Figure 4.

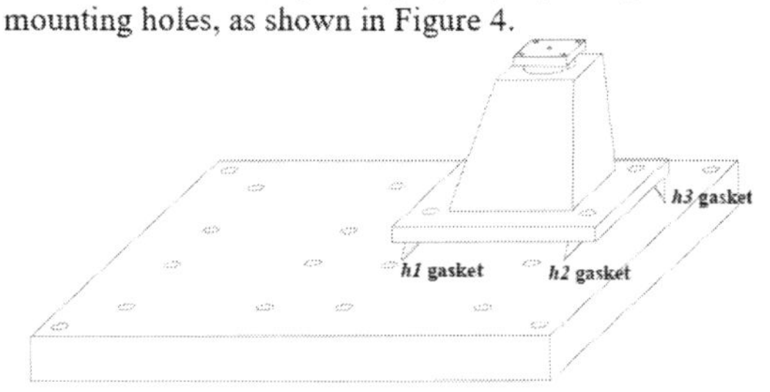

Figure 4. Pressing seat installation and adjustment diagram

Since the lower pressing seat of each pressing mechanism is flexible and adjustable, it is difficult to ensure that the mating surfaces of the upper and lower pressing seats of each pressing mechanism can meet the matching requirements during the assembly of the whole equipment, and it is easy to have partial or most of the mating surfaces pressing mechanism offset, skew, clearance and other situations, which eventually lead to the failure of the matching, stress concentration and other problems, and ultimately affect assembly quality.

In general, multi-point pressure-type equipment is mostly multi-faceted and is usually fixed to a specific tooling (called a simulated wall) before delivery. All the pressing mechanisms on the tooling are consistent with the cabin and the rigidity is good, which can well maintain the posture state of the device. The installation state of the equipment on the simulated wall is the best installation state, and the fitting tolerance of each pressing device is satisfactory. According to the requirements, all the pressing mechanisms (lower pressing seats) on the simulated wall must be removed and installed on the cabin, and then the pressing mechanism on the cabin is adjusted during the loading process. The posture state is consistent with the tooling state, thereby realizing the installation of the device on the cabin.

3 Transfer scheme of benchmark based on mounting holes

For ease of description, the multi-point pressure-type large equipment for loading is classified as a docking assembly, and the mounting surface of the cabin is referred to as a reference assembly, and the simulated wall is referred to as a tooling.

In the digital assembly, the transfer of the reference is carried out step by step from the actual object. The reference coordinate system quoted in the measurement is an important part of the reference transmission, which directly affects the model posture and state changes observed by the operator in the three-dimensional computer space. In the process of benchmark transmission, it is necessary to ensure the consistency of the benchmark transmission and reduce the error of the reference transmission, so as to effectively reduce the deviation of the attitude caused by the deviation of the coordinate system.

In the process of digital assembly of docking assembly on spacecraft, the transfer of datum usually involves three coordinate systems: design coordinate system, tooling coordinate system and datum assembly coordinate system (assembly coordinate system).

The design coordinate system is the reference coordinate system used in the design process of the equipment. Usually, one of the key point of the product is taken as the origin of the coordinate system, and the axis of symmetry or the central axis is taken as the axial direction, and the horizontal plane is parallel with the axis to establish another axial direction. The right hand rule establishes the third axis. The tooling coordinate system is the coordinate system represented by the specially processed measuring hole position and its calibration coordinates after the tooling design is completed and processed. In general, the design coordinate system and tooling coordinate system of docking assembly have been determined in the stage of design and production department assembly, and relevant measurement work has been completed by the designer. In the loading phase, the tooling coordinate hole is marked on the tooling. The coordinate system consistent with the design coordinate system can be quickly obtained on the tooling, and only the coordinate system of the reference assembly needs to be determined later.

However, in most cases, the reference assembly is limited by its own functional and structural requirements, and it is not allowed to directly manufacture the dedicated reference hole corresponding to the tooling as the established value of the reference coordinate system, so there is no reference hole position with the same coordinate system on the tooling, which make it impossible to establish a coordinate system consistent with the tooling coordinate system during assembly, so the transfer of the reference cannot be smoothly performed.

In order to solve the above problems, considering that the mounting holes of the pressing mechanism are provided on the tooling and the reference assembly, and the hole position layout is consistent with the topological relationship, the installation hole of the pressing mechanism at some specific positions is taken as the reference hole position established in the coordinate system. So the reference coordinate system on the tooling can be transferred to the reference assembly (i.e. the cabin) by this reference transfer method. The reference system transfer process is shown in Figure 5, and the partial compression mechanism mounting holes in the figure have been omitted.

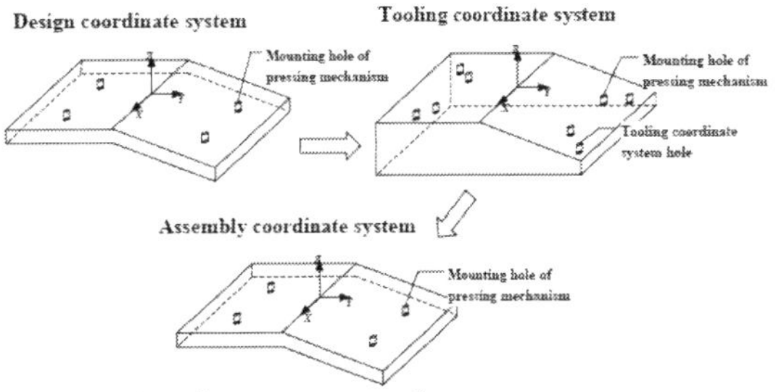

Figure 5. Coordinate system transfer process

According to the existing working conditions, there are three options for selecting the measured value of the mounting hole of the pressing mechanism as follows:

- Select the design theoretical value given by the design coordinate system and the coordinate system transfer route is from the design coordinate system to the assembly coordinate system;
- Select the measured value obtained by the designer according to the physical measurement and verification. This scheme is an improvement of the first scheme;
- Select the measured value of the corresponding mounting hole on the tooling under the current working condition environment. The transfer route of coordinate system is from the design coordinate system to the tooling coordinate system and then to the assembly coordinate system, which requires two times of datum transfer process.

Because the docking assembly is in the best fit with the tooling after undergoing transportation, hoisting and pre-testing, it is impossible to objectively agree with the theoretical value of the design. Therefore, only the mounting state of the pressing mechanism on the tooling can accurately reflect the posture of the pressing mechanism on the docking assembly, and the coordinate system on the tooling can best reflect the state of the docking assembly. Therefore, the third option is the best choice.

In the actual assembly, after the butt assembly is separated from the tooling, the lower pressing seat is taken out one by one and restored to the reference assembly according to the tooling, so that the optimal reference assembly can be matched with the docking assembly. In selecting the mounting hole of the pressing mechanism, a total of 16 mounting holes covering the outer edges of the main mounting surface in four pressing mechanisms can be selected as the coordinate system reference point.

4 Compensation calculation method

According to the structural characteristics of the lower pressing seat of the pressing mechanism, the measuring aid is used to measure the posture of the measuring device. The measuring aid has 5 measurable hole positions, wherein the central hole position is the spare measuring hole position, and only 4 holes are needed for measurement, as shown in Figure 6.

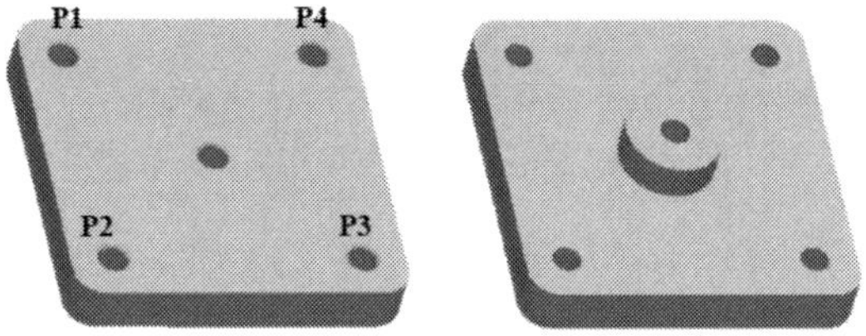

(a) Front (b) Back

Figure 6. Schematic diagram of the measuring aid

According to formula (4), the position and normal vector of the center point of the measuring aid can be calculated, and the same measuring aid can be used to measure the pressing mechanism to avoid systematic errors caused by using different measuring aids.

$$\begin{cases} \overrightarrow{O_i} = \frac{\overrightarrow{(P_3-P_1)}\times\overrightarrow{(P_4-P_2)}}{\left|\overrightarrow{(P_3-P_1)\times(P_4-P_2)}\right|} \\ P_i = \frac{1}{4}\sum_{j=1}^{4} P_j - L\cdot\overrightarrow{O_i} \end{cases} \tag{4}$$

We can restore the target attitude parameter on the reference assembly, which is measured by the measuring aid in the tooling coordinate system. If we calculate the amount of gasket and the amount of translation at the installation hole according to the geometric dimension of the compression mechanism, we need to consider the installation direction of the compression mechanism and other conditions, which will make the calculation process very complicated. Therefore, we can use the method based on the change of rigid body to calculate, so that the four mounting holes of the pressing mechanism and the position and direction parameters of the measuring auxiliary center point calculated by formula (4) form a pyramid. Since the rotation of the measuring aid does not affect the position of the center point, the pyramid always remains relatively unchanged, as shown in Figure 7.

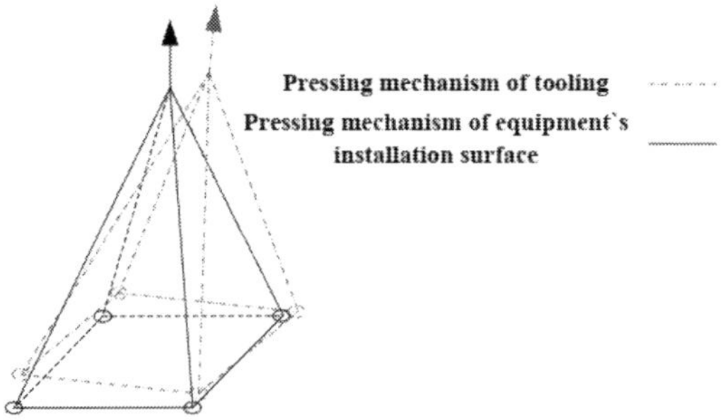

Figure 7. Pressing mechanism adjustment calculation diagram

Therefore, the calculation of the variation includes two parts, the direction vector rotation and the positional degree translation. The direction vector is solved by the Rodrigue rotation matrix, and the translation is solved by vector subtraction, as shown in equations (5) and (6).

$$\overrightarrow{O_i^D} = R\overrightarrow{O_i^T} \tag{5}$$

$$R = -\left(\cos(\theta)\vec{I} + (1-\cos(\theta))rr^T + \sin(\theta)\begin{bmatrix} 0 & -r_z & -r_y \\ r_x & 0 & -r_z \\ -r_y & r_x & 0 \end{bmatrix}\right) \tag{6}$$

In the formula, $\overrightarrow{O_i^D}$ and $\overrightarrow{O_i^T}$ are the measured direction vectors on the reference assembly and tooling respectively. θ is the angle between the two, and r is the unit obtained by the cross-multiplication vector. After calculating the rotation matrix, rotate and transform the mounting hole of each pressing mechanism to obtain the new position of the point, then calculate the distance between the new position and the original position on the direction vector, and subtract the minimum gasket thickness together, so as to obtain the gasket thickness value of each mounting hole to compensate the rotation deviation. Calculate the position of the apex of the pyramid at this time, and subtract it from the position parameters measured by the tooling. The obtained amount is the translation amount. By projecting the calculation amount on the plane of the direction vector direction and the vertical direction vector respectively, the gap compensation amount and the translation compensation amount caused by the gap can be calculated.

5 Multi-point compression assembly test

The above theory is verified by the installation of a large multi-point pressure-type equipment on a manned spacecraft. The assembly process is shown in Figure 8.

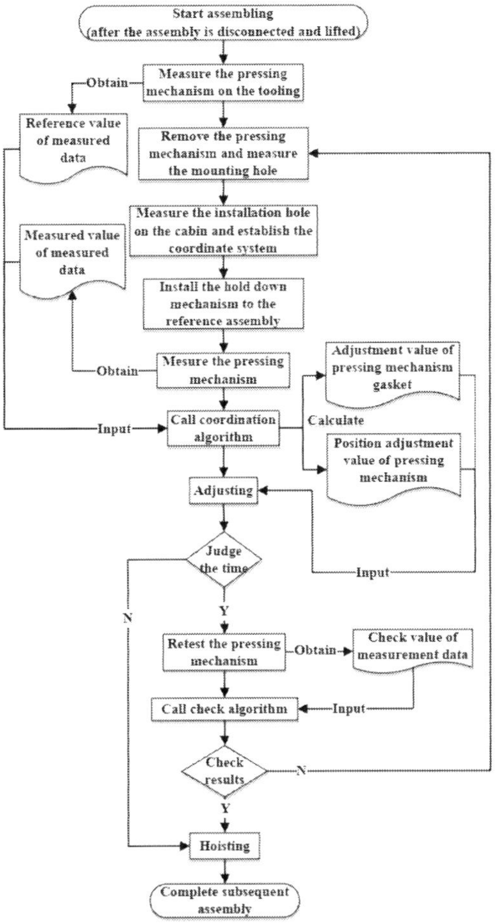

Figure 8. Multi-point pressing mechanism assembly process

Firstly, the tooling coordinate system is established according to the coordinate system calibration hole which is on the tooling, and the parameters of the pressing mechanism on the tooling are measured by the measuring aid. Then dismantle the pressing mechanism on the tooling, and measure the installation hole of the coordinate system to be established on the cabin. And Then measure the corresponding installation holes on the cabin and establish the coordinate system. After all the pressing mechanisms are installed, we can measure all the pressing mechanisms and calculate the compensation amount according to the calculation method of the adjustment amount, and adjust the position and posture one by one according to the compensation amount. Finally, retest and inspection are carried out.

For this assembly process, we developed the measurement and data analysis of multi-point pressing assembly coordination software to assist the assembly and adjustment process, and integrated the laser tracker measurement, data storage and management, data analysis, three-dimensional display, online assembly and adjustment oriented to the laser tracker and other functional modules. Part of the software operation interface is shown in Figure 9. The online adjustment is to ensure that the spatial position of the pressing mechanism can be instantly reacted. The tracking measurement mode of laser tracker is adopted, and the target ball is placed on the measuring aids to compare its deviation with the target value, so as to analyze whether the pressing mechanism is installed and adjusted successfully in real time.

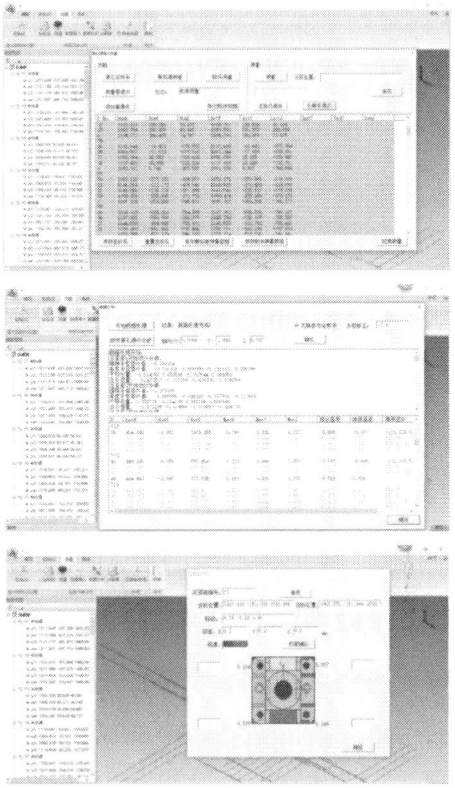

Figure 9. Multi-point compression coordination software interface

Table 1 shows the adjustments calculated by the software after the measurement of the equipment pressing mechanism. The equipment has a total of 41 sets of pressing mechanisms. Because of the huge amount of data, Table 1 only lists the adjustment of the pressing seats from Y1 to Y4. In the table, the compensation value of the gasket of the four hole positions of each pressing seat is divided into the theoretical value and the correction value. Considering that the compensation value of the hole position gasket of the pressing seat must be positive (that is, the pressing seat can only be padded upward, not downward), so the theoretical value is modified as a whole according to the topological relationship of the equipment pressing seat. The minimum negative value is modified to 0, and the rest values are compensated accordingly by using the software. So the corrected gasket compensation value can be obtained finally.

The X and Z directions defined in the table are respectively the two directions of the pressing seat along the mounting plane, which can be adjusted online by software until the requirements are met. The Y direction is the normal direction of the mounting surface of the pressing seat, and its translation adjustment is ultimately based on the gasket compensation value of the mounting hole, so it is not the basis for adjusting the posture of the pressing seat itself, which can be ignored.

Table 1. Measurement results before the coordination of the pressing mechanism (Partial data) (Unit: mm)

No. of pressing mechanism	Translation value in X direction	Translation value in Y direction	Translation value in Z direction	Hole position of lower pressing seat	Theoretical value of gasket compensation	Correction value of gasket compensation
Y1	-0.4476	-0.2721	0.2916	1	-0.5848	0.0960
				2	-0.6696	0.0111
				3	-0.6879	0.0072
				4	-0.6026	0.0781
Y2	0.4961	-0.2055	0.2205	1	-0.3141	0.3667
				2	-0.5083	0.1725
				3	-0.5758	0.1050
				4	-0.3828	0.2980
Y3	0.3536	0	-0.1093	1	0.1426	0.8234
				2	0.2028	0.8836
				3	0.2236	0.9043
				4	0.1633	0.8441
Y4	0.166	-0.0707	-0.0757	1	-0.2003	0.4805
				2	-0.1257	0.5550
				3	-0.0760	0.6047
				4	-0.1511	0.5296
......	......	......	......	......	......	......

Table 2. Retest results after compression of the pressing mechanism (Partial data) (Unit: mm)

No. of pressing mechanism	Translation value in X direction	Translation value in Y direction	Translation value in Z direction	Hole position of lower pressing seat	Theoretical value of gasket compensation	Correction value of gasket compensation
Y1	-0.014	0.0536	-0.0636	1	-0.4202	0.0899
				2	-0.4101	0.1000
				3	-0.4262	0.0839
				4	-0.4363	0.0738
Y2	-0.0155	-0.0082	0.0665	1	-0.4661	0.044
				2	-0.458	0.0521
				3	-0.4359	0.0742
				4	-0.4463	0.0638
Y3	0.1332	0.0001	0.1701	1	-0.4624	0.0477
				2	-0.4973	0.0128
				3	-0.4762	0.0339
				4	-0.4412	0.0689
Y4	-0.1068	0.0267	0.0740	1	-0.5071	0.003
				2	-0.4609	0.0492
				3	-0.452	0.0581
				4	-0.4981	0.012
......	......	......	......	......	......	......

According to the calculated translation value and the gasket compensation value, after all the lower pressing seats of the equipment is adjusted in position, the measuring aid and the laser tracker are used again to retest, and the retest results are shown in Table 2.

It can be seen from Table 2 that the positional deviation is controlled within 0.2 mm according to the previously adjusted compensation value, and the height deviation is controlled within 0.1mm to meet the assembly requirements of the equipment. According to statistics, the entire installation cycle of the equipment is about 12 hours. Compared with the previous traditional adjustment time (24 hours), the efficiency is increased by 50% and the quality of installation and adjustment becomes more reliable by using this method.

6 Conclusion

Aiming at the assembly process of multi-point pressure-type large-scale equipment on spacecraft, a method of multi-point compact assembly and adjustment based on digital measurement is proposed in this paper, in which the geometric structure of multi-point compact is analyzed, and the mathematical model of geometric constraints is established, and the matching relationship of the whole assembly is also analyzed. In view of the lack of positioning datum on the cabin, this paper analyzes the datum assembly form in the assembly process, and establishes the datum transfer and coordinate system establishment scheme, and then gives the calculation method of gasket compensation. The method proposed in this paper is helpful to improve the assembly and adjustment of multi-point pressing type, ensure the assembly quality and improve the assembly efficiency, which can provide reference for other similar space equipment by using digital assembly methods.

References

1. Binglong W, Daokui Q, Fang X. Industrial Robot High Precision Peg-in-hole Assembly based on Hybrid Force/Position Control. J. *Journal of Zhejiang University* (Engineering Science), **52**, 379(2018)
2. Mei Z, Maropoulos P G. Review of the Application of Flexible, Measurement-assisted Assembly Technology in Aircraft Manufacturing. J. Proceedings of the Institution of Mechanical Engineers, Part B: *Journal of Engineering Manufacture*, **228**, 1185(2014)
3. Yuan L, Li Z, Yanzhong W. An Optimal Method of Posture Adjustment in Aircraft Fuselage Joining Assembly with Engineering Constraints. J. *Chinese Journal of Aeronautics*, **30**, 2016(2017)
4. Junxiu L, Zongquan D, Honghao Y. Research on locked point arrangement method of spatial mechanical arm. J. *Jouenal of Machine Design*, **29**, 85(2012)
5. Zongquan D, Binghui W, Xiang W. Design and analysis of anew style lock-unlock mechanism for spaceborne appendages. C. *Proceedings of the International conference on Advanced Design and manufacture*, 117(2006)
6. Jun W. Research on Large Load and Distributed Locking-releasing Mechanism of Space Manipulator. D. Harbin Institute of Technology, (2014)
7. Graham G, External R, Savi S, et al. Canada and the international space station program: overview and status. J. *Acta Astronautica*, **51**, 59(2002)
8. Heemskerk C J M, Schoonejans P H M. Overview of software engineering applications in the european robotic arm. C. *Proceedings of the NASA 97 Conference on "Data System in Aerospace"*, 317(1997)
9. Huifang Z, Zongquan D, Xiang W, et al. Kinematic Analysis of releasing process of Locking- releasing Mechanism for Robotic Arm. J. *Mechanical Engineer*, 49(2006)

Research and Implementation of an Automatic Reset Force of a Force Feedback Handle

Xin An Qiu[1,a], Shi Jia Wang[1] and Dong Tao Ma[1]

[1] Lanzhou Institute of Physics, Lanzhou 730000, China

Abstract. Take the force feedback handle applied to the teleoperation of space robotic arm as a requirement. In order to improve users' experience, we studied the automatic reset force of the handle. This paper proposes a spring-damping model and applies it to the torque output of the motor to achieve a good reset of the handle, which is a new development of the application field of the automatic reset force model of the force feedback device. The experiment shows that the automatic reset force model has high accuracy when the handle returns to zero. In addition, through dynamic and reasonable adjustment of the stiffness coefficient and damping coefficient, it can meet the needs of different users for the automatic reset force of the force feedback handle.

1 Introduction

In recent years, China 's space industry has developed vigorously. According to the "three-step" development strategy, China will build its own manned space station around 2020. Space robotic arm, which is one of the three key technologies of the space station, will be used in the construction and on-orbit maintenance of our space station[1].

With the continuous deepening of the construction of China 's space station, astronauts can complete the capture docking and indexing of space segments, assisted replacement of out-of-vehicle equipment, self-exit activities, and platform load care by operating robotic arms. Its complicated operation flow and the special environment of the space require the robot arm operating system to have better multi-dimensional interaction capabilities in order to better assist astronauts to complete the robot arm operation processing.

As a human-computer interface of force perception[2], force interaction device can complete the information collection of operator's operation action, realize the control of the operated object, and feedback the state of the operated object to the operator in the form of force, so as to improve the "sense of presence" and "sense of transparency" of its operation [3-5]. At present, teleoperation technology based on force sense interaction is widely used in robot systems working in special environments, such as space, danger, radiation, high temperature, toxicity or seabed environment[4,6]. On the basis of vision and hearing, it increases the way of force information feedback, which greatly improves the efficiency, accuracy and safety of the operating system. It is of great significance to apply it to the master-slave control of the space station manipulator[7,8].

Based on the above requirements, a force feedback handle based on 3-RRR spherical parallel mechanism is designed. The force feedback handle shall have the function of automatically returning to the zero position. The operator shall overcome the automatic reset force to operate. After the operation, the automatic reset force will return the handle to the zero position. The general handle is installed with spring in the mechanism to form an automatic reset force, but it has many disadvantages such as spring deformation and difficulty in self-regulation, so the user experience is poor. In the parallel mechanism, the motor is used to provide the torque, and the return force of the spring is simulated by the servo motor on three joints.

2 Research on model of automatic reset force of handle

Generally, the control handle needs to have the force of returning to the zero position. When the operator carries out remote control operation, it is necessary to pull the handle to operate against the recovery force. After the operation, the hand leaves the handle, and the handle can automatically return to the zero position. Considering that the working space of the handle is ± 40 ° around each axis, in order to have no singular point and member interference in the selected working space, and to have better flexibility index and force transmission performance index, the zero position (initial position) of the mechanism is defined as shown in Figure 1. That is the reclosing position of the axis connecting the moving platform and the connecting rod, and the axis connecting the fixed platform and the connecting rod of the adjacent moving chain.

[a] Corresponding author: qiuxinan502@163.com

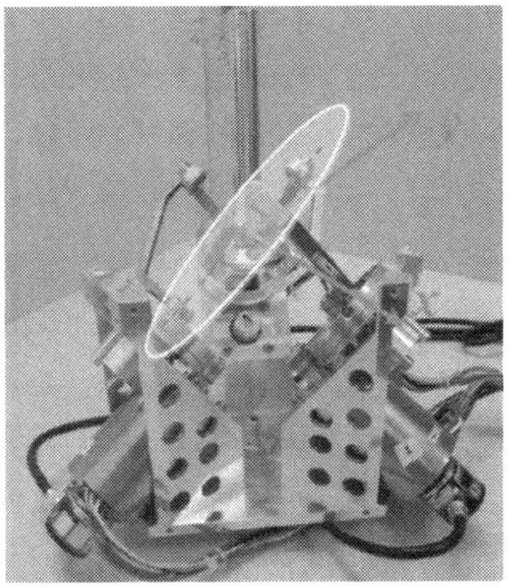

Figure 1. Definition of mechanism zero position

The automatic reset force of the handle is related to the movement of the end. The displacement of the joint angle is detected by the encoder installed at each joint of the handle. The torque required to be output by each driving motor is calculated according to the corresponding automatic reset force calculation method, and the corresponding electric current output by the driving motor is controlled, so as to realize the real-time automatic reset force output of the handle.

The common control handle relies on the return spring to realize the automatic return to zero position. The force feedback handle designed in this paper relies on the motor on three joints to provide power to complete the return to zero position, so it needs to simulate the return force of the spring through the servo motor.

2.1 Motor simulation of spring

In order to obtain more physical recovery force, the coordinate system is established at the end handle of the mechanism, and the zero-position attitude is defined as the fixed coordinate system O-x0y0z0, the real-time attitude of the handle is the dynamic coordinate system O-xyz, as shown in Figure 2. In the figure, i, j, k represents the unit vector of the dynamic coordinate system O-xyz, i0, j0, k0 represents the unit vector of the fixed coordinate system O-x0y0z0.

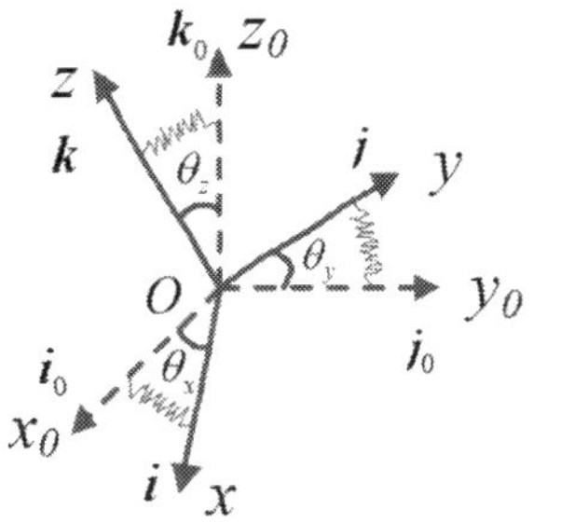

Figure 2. Definition of the included angle between the actual end position of the handle and the zero position

Let θ_x, θ_y, θ_z be the angle between the unit vectors i0 and i, j0 and j, and k0 and k, then

$$\begin{cases} \cos\theta_x = \boldsymbol{i}^T \bullet \boldsymbol{i}_0 \\ \cos\theta_y = \boldsymbol{j}^T \bullet \boldsymbol{j}_0 \\ \cos\theta_z = \boldsymbol{k}^T \bullet \boldsymbol{k}_0 \end{cases} \tag{1}$$

Assuming that a spring (or torsion spring) is installed between each coordinate axis of the coordinate system O-x0y0z0 to the dynamic coordinate system O-xyz the angle between the return moment and the coordinate axis is proportional. Set the return moment between the x axis and x0 axis as Tyz, the return moment between the y axis and y0 axis is Txz, and the return moment between the z axis and the z0 axis be Txy, The stiffness coefficient of each spring is k_{yz}, k_{xz}, k_{xy}, then

$$\begin{cases} T_{yz} = k_{yz}\theta_x \\ T_{xz} = k_{xz}\theta_y \\ T_{xy} = k_{xy}\theta_z \end{cases} \tag{2}$$

The axis direction of T_{yz}, T_{xz}, T_{xy} is set to $\mathbf{e}_{yz}$, $\mathbf{e}_{xz}$, $\mathbf{e}_{xy}$, then

$$\begin{cases} \boldsymbol{e}_{yz} = \boldsymbol{i}_0 \times \boldsymbol{i} \\ \boldsymbol{e}_{xz} = \boldsymbol{j}_0 \times \boldsymbol{j} \\ \boldsymbol{e}_{xy} = \boldsymbol{k}_0 \times \boldsymbol{k} \end{cases} \tag{3}$$

Define the torque vector as $\boldsymbol{T}_{yz}$, $\boldsymbol{T}_{xz}$, $\boldsymbol{T}_{xy}$, then

$$\begin{cases} \boldsymbol{T}_{yz} = T_{yz}\boldsymbol{e}_{yz} \\ \boldsymbol{T}_{xz} = T_{xz}\boldsymbol{e}_{xz} \\ \boldsymbol{T}_{xy} = T_{xy}\boldsymbol{e}_{xy} \end{cases} \tag{4}$$

The resultant torque vector of the automatic reset force at the end of the handle is

$$\boldsymbol{T}_r = \boldsymbol{T}_{yz} + \boldsymbol{T}_{xz} + \boldsymbol{T}_{xy} \tag{5}$$

Let the angular velocity of the end of the mechanism be $\boldsymbol{\omega}$, and the angular velocity of the driving joint be $\dot{\boldsymbol{s}}$. From reference [5], the speed relationship of the 3-RRR spherical parallel mechanism can be described as:

$$\boldsymbol{A}\boldsymbol{\omega} = \boldsymbol{B}\dot{\boldsymbol{s}} \tag{6}$$

Among them, $\boldsymbol{A}$ is the first type of Jacobian matrix, and $\boldsymbol{B}$ is the second type of Jacobian matrix

$$\boldsymbol{A} = \begin{bmatrix} (\boldsymbol{w}_1 \times \boldsymbol{v}_1)^T \\ (\boldsymbol{w}_2 \times \boldsymbol{v}_2)^T \\ (\boldsymbol{w}_3 \times \boldsymbol{v}_3)^T \end{bmatrix} \tag{7}$$

$$\boldsymbol{B} = diag(\boldsymbol{u}_i \times \boldsymbol{w}_i \bullet \boldsymbol{v}_i) \quad i = 1,2,3; \tag{8}$$

$\boldsymbol{u}_i$, $\boldsymbol{w}_i$, $\boldsymbol{v}_i$ $(i = 1,2,3)$ are the direction of the three rotating auxiliary axes (drive shafts, hinges of connecting rods and connecting rods, hinges of connecting rods and moving platforms)of each moving branch chain.

Let $\boldsymbol{J}$ be the speed mapping relationship between the input and output of the mechanism ($\dot{\boldsymbol{s}} = \boldsymbol{J\omega}$), namely the speed Jacobian matrix, then

$$\boldsymbol{J} = \boldsymbol{B}^{-1}\boldsymbol{A} \tag{9}$$

In the control, it is necessary to make use of the Euler angle to differentiate the time, so the transformation matrix $\boldsymbol{R}$ is introduced.

$$\omega = \boldsymbol{R}\dot{\boldsymbol{t}} \tag{10}$$

Where t is the Euler angle vector, $\boldsymbol{t} = [\phi, \theta, \psi]^T$.

$$\boldsymbol{R} = \begin{bmatrix} 0 & \cos\phi & -\cos\theta\sin\phi \\ 0 & \sin\phi & \cos\theta\cos\phi \\ 1 & 0 & \sin\theta \end{bmatrix} \tag{11}$$

Thus, it can be obtained

$$\dot{s} = \boldsymbol{JR}\dot{\boldsymbol{t}} \tag{12}$$

Let the driving torque of the drive motor be $\boldsymbol{\tau}$,
According to the principle of virtual work, there are

$$\boldsymbol{T}_r^T\dot{\boldsymbol{t}} = \boldsymbol{\tau}^T\dot{\boldsymbol{s}} \tag{13}$$

Substituting equation (12) into (13) gives

$$\boldsymbol{\tau} = (\boldsymbol{JR})^{-T}\boldsymbol{T}_r \tag{14}$$

In the formula, $\boldsymbol{JR}$ is an extended Jacobian matrix.

2.2. Study of damping force

The spring model can be used to simulate the return force of the spring. In the actual debugging, the spring stiffness coefficient must be set carefully. If the spring stiffness coefficient is too small, the handle will lose power when it returns to the zero position, causing a large zero -return error. If the spring stiffness coefficient is too large, the automatic reset force is too large, and the handle is easy to cause oscillation due to the excessively fast restoring speed. In order to ensure that the handle has a high return-to-zero accuracy, and to minimize the back and forth oscillation phenomenon, a damping force needs to be introduced into the mechanism.

The damping force is a resistive moment that is proportional to the angular velocity, and its direction is opposite to the movement direction. Damping helps reduce the transient impact received by the mechanical system, hinders the structure from transmitting vibrations, and is beneficial to the stability of the device. In practical mechanical systems, damping is always present because of friction. In the automatic reset force control of the force feedback handle, increasing the damping term helps to increase the reliability of the zero return and reduce vibration.

The most commonly used damping model is linear viscous damping. The magnitude of the damping force is proportional to the speed and opposite to the direction. For the force feedback devices, when the spring stiffness coefficient is too large, it often causes oscillation. Adding a damping term can significantly improve this situation, so a damping term should be added to the force control model.

Let the damping force be T_{dmp}, then

$$T_{dmp} = B\omega \tag{15}$$

Here B is the damping coefficient and ω is the joint angular velocity. For discrete systems,

$$\omega = (\theta_t - \theta_{t-1})/t_{sample} \tag{16}$$

In the formula, θ_t refers to the angular displacement at time t; θ_{t-1} refers to the angular displacement at time t-1, t_{sample} is the sampling time interval.

For force feedback control systems, higher sampling frequency is usually required. t_{sample} is generally set to 10ms or less. In discrete sampling, the calculation of speed is not continuous and it is prone to sudden changes in speed. This requires the selection of a suitable damping coefficient B. If B is too large, it may cause instability due to discontinuities; if B is too small, the effect of the damping term is not obvious, which is also not conducive to the stability of the control.

The calculation of the damping term is directly calculated at each drive joint. The angular displacement in each sampling period is collected by the encoder, and after the angular velocity is calculated, we can obtain the damping moment: T_{dmp}, and then superimpose it with the automatic reset force calculated by the previous inverse solution and output it to the motor.

3 Achieve

The handle automatic reset force is calculated using the spring-damping model, as shown in Figure 3.

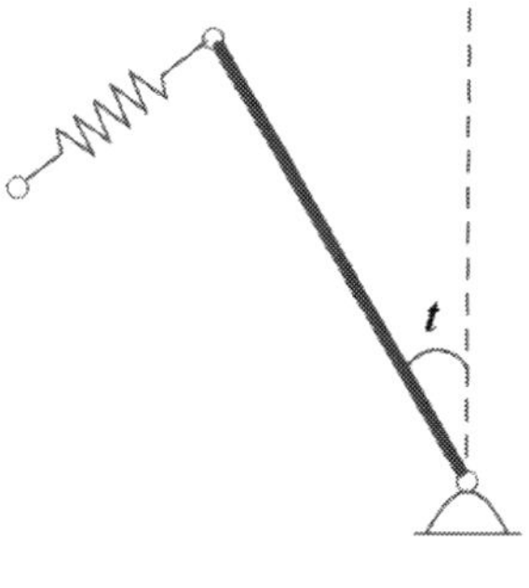

Figure 3. Spring-damped model

The formula for calculating the automatic reset force can be expressed as:

$$\mathrm{F} = \mathrm{Kt} + \mathrm{B}\dot{\mathrm{t}}$$

Among them, $\boldsymbol{t} = [\phi, \theta, \psi]^{\mathrm{T}}$ is the Euler angle at the end, $\boldsymbol{K}$ is the stiffness matrix, $\boldsymbol{B}$ is the damping matrix.

Suppose the Euler angle obtained by iterative calculation is φ, θ, ψ , then the calculation process of the spring force is shown in figure 4, the calculation process of the damping force is shown in Figure 5.

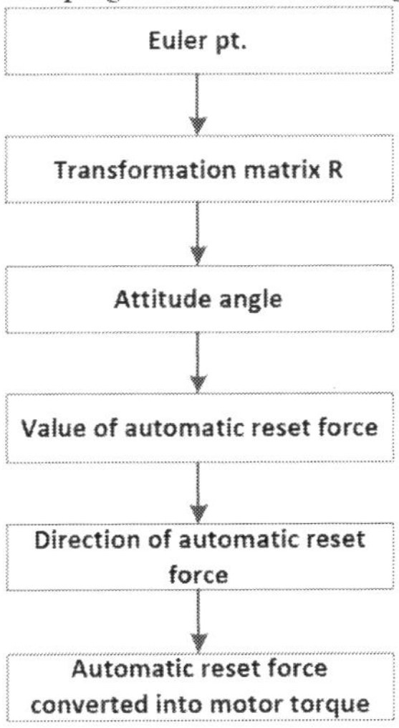

Figure 4. Calculation process of automatic reset force

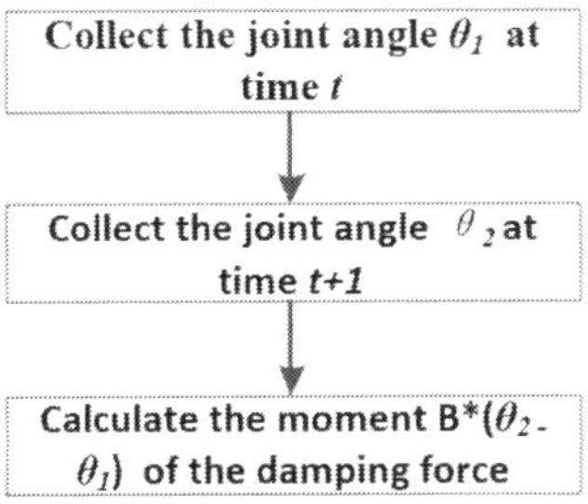

Figure 5. Calculation process of damping force

4 Experimental verification

The automatic reset force of the handle is the center force of the handle returning to zero, which is reflected as a force that always points to the center zero during the operation. When the operation is completed and the operator's hand leaves the handle, the handle mechanism automatically returns to the zero position under the action of the automatic reset force. The magnitude of the automatic reset force and the accuracy of returning to zero are indicators for evaluating whether the law of returning force is appropriate. A larger returning force will increase the resistance during operation, which will easily cause the operator's fatigue. A smaller returning force will make the return to zero slowly and zero-return accuracy is poor. Reasonably adjusting the parameter stiffness coefficient ***K*** and damping coefficient ***B*** can change the magnitude of return force and zero return error. The automatic reset force experiment is shown in Figure 6.

(1) the effect of damping coefficient on the stability of the automatic reset of the handle

Set the constant spring stiffness coefficient as $\boldsymbol{K}$=[8Nmm/° 3Nmm/° 3Nmm/°]^T, change the damping coefficient from 0.5 Nmm/rpm to 1.6Nmm/rpm， Set the constant spring stiffness coefficient as K = 1, change the damping coefficient from 0.5 Nmm / rpm to 1.6Nmm / rpm, perform an automatic reset experiment on the handle, record the angle difference between the Z-axis direction of the handle and the zero position and the recovery torque of motor output to handle, and the experimental results are shown in figure 7.

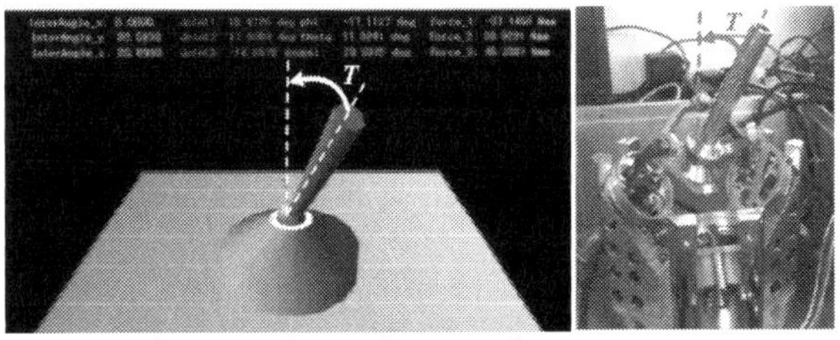

Figure 6. The automatic reset force experiment

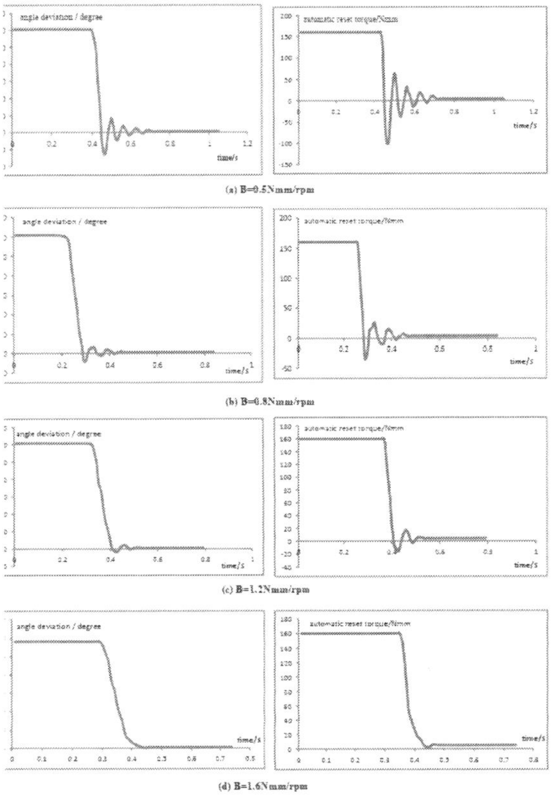

Figure 7. Effect of damping coefficient on the return-to-zero process

From the above comparison, it can be seen that when the damping coefficient B = 0.5Nmm/rpm, the handle will oscillate obviously around the zero position of automatic reset. When the damping coefficient is increased, the back and forth oscillation phenomenon gradually weaken, and the handle will soon stop at the zero position. When the damping coefficient is increased to B=1.6Nmm/rpm, the oscillation phenomenon disappears, and the handle can automatically return to the zero position and stop. The above phenomenon shows that the damping term is of great significance to the improvement of the zero-return stability of the handle and reasonably setting the damping coefficient can control the handle to return to zero automatically and stably.

(2) Effect of spring stiffness coefficient on zero return accuracy

Spring stiffness coefficient $\boldsymbol{K}$=[k_{xy},k_{yz},k_{xz}]^T, The latter two components are the restoring moments when the handle is rotated around its Z-axis. Because the force arm perpendicular to the Z-axis is short, the moment in this direction should be set smaller, or it will make operation difficult. Here only study the zero return situation where the handle deviates from the vertical direction, so only change the value of k_{xy} from 6Nmm/° to 14Nmm/°, investigate the angular error of the handle returning to zero (represented by the mean error of the ZXY Euler angle), and the experimental results are shown in the figure.

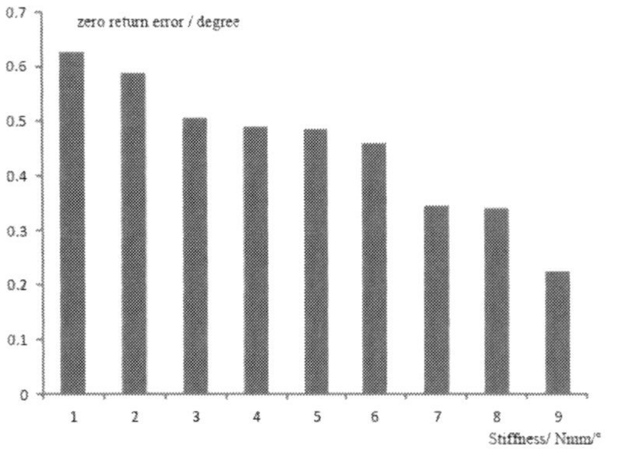

Figure 8. Effect of changing spring stiffness coefficient on return-to-zero accuracy

The above experiments show that increasing the spring stiffness coefficient is conducive to improving the zero-return accuracy of the handle. However, a large spring stiffness will cause overshoot to zero and back and forth oscillation, which is not conducive to the stability of zero return. Therefore, a suitable spring stiffness should be selected in combination with the damping.

5 Conclusion

Aiming at the requirement that the force feedback handle needs to return to zero automatically, this paper studies a spring-damping model and applies its model to the torque output of the motor to achieve a reliable reset of the handle. The experiment shows that the reasonable setting of the damping coefficient can control the handle to return to the zero position stably and reliably, and the reasonable stiffness coefficient can balance the accuracy and stability of returning to zero. By dynamically and reasonably adjusting the stiffness coefficient and the damping coefficient, it can meet the needs of different users for the automatic reset force of the force feedback handle.

References

1. LI Daming, RAO Wei, HU Chengwei. Key Technology Review of the Research on the Space Station Manipulator [J]. Manned Spaceflight, **20**, 3 (2014) : 238-242.
2. Chipalkatty R, Daepp H, Egerstedt M, et al. Human-in-the-loop: MPC for shared control of a quadruped rescue robot[C]. IEEE/RSJ International Conference on Intelligent Robots and Systems. IEEE, (2011) : 4556-4561.
3. Preusche C, Hirzinger G. Haptics in telerobotics[J]. The Visual Computer, **23**, 4 (2007): 273-284.
4. Goertz R C. Mechanical master-slave manipulator[J]. Nucleonics (US) Ceased publication, (1954).
5. Munjal Desai M, Tsui K M, Yanco H A, et al. Essential features of telepresence robots[C]. IEEE Conference on Technologies for Practical Robot Applications, (2011) :15-20
6. SONG Aiguo. Force Telepresence Telerobot : Review of the history and development[J]. Journal of Nanjing University of Information Science and Technology: Natural Science Edition，**5**, 1 (2013) : 1-19
7. Xiao dongg ZHANG Xiaodong. Research on Control Strategy of Space Flexible Manipulator[D]. Bei Jing: Beijing University of Posts and Telecommunications, (2008).
8. Deng Le, Zhao Ding-xuan, Tang Xin-xing. Research on key technologies of controller for teleoperation force reflecting telepresence [J]. Journal of Engineering Design, **12**, 3 (2005) :159-162.

Structure Analysis and Design of Automobile plastic clutch pump body based on friction welding

Jing Sheng[1,2,a], Aamir Sohail[2], Mengguang Wang [2] and Zhimin Wang[2]

[1] *Fujian Key Laboratory of Advanced Design and Manufacture for Bus$Coach, Xiamen University of Technology, 361024 Fujian Xiamen, China*
[2] *Fujian Collaborative Innovation Center for R&D of Coach and Special Vehicle, Xiamen University of Technology, 361024 Fujian Xiamen, China*

Abstract. In order to realize the need for lightweight automobiles through replacing steel with plastics, the research and development of the plastic clutch pump body based on the friction welding was carried out. For the clutch pump body connected by friction welding process between the upper pump body and the lower pump body, the technical requirements of pressure 14 MPa and durability (high temperature 7.0×10^4 times, room temperature 7.0×10^5) are required. The structure type of the upper and lower pump bodies of the end face welding type was proposed. Through the static analysis of the pump body and weld and the mechanical analysis under the working condition, the structure of the clutch pump body (upper and lower pump body) was determined. According to the established welding process, the pressure of the clutch pump body is more than 15 MPa, and the number of high-temperature durable circulation and the number of room temperature durable circulation also reached 7.2×10^4 and 7.3×10^5 times respectively. The results show that the structural design of a clutch pump body meets the design requirements.
Keywords: Plastic clutch pump body; Rotating friction welding; Welding surface structure; Finite element analysis; Durability

1 Introduction

As plastic parts can meet the requirements of lightweight and satisfy the safety and environmental protection standards, automobile plastic parts emerge at the right moment. "Replacing Steel with Plastics" has become the trend of parts development in the automotive industry. The plastic clutch master pump (Fig. 1a) is an important component of the automobile clutch control system. Its function is to convert the force input by the clutch pedal into hydraulic pressure and output it to the clutch slave cylinder, which converts the hydraulic pressure into thrust and pushes the clutch release bearing to realize the clutch separation [1]. As the key part of the clutch master cylinder, the clutch pump body (Fig. 1b) has some problems for example air leakage and short service life, so the plastic clutch master cylinder has not been widely used up to now. Therefore, in order to provide a reference for its design, it is necessary to analyze the structure of the pump body of the plastic clutch considering the welding process.

Plastic processing technology expands the manufacturing process of auto parts [2-4]. In rotary welding, Tappe et al. [5] analyzed the effect of welding parameters and material properties on the temperature of the welding zone in plastic rotary welding and discussed the influence of welding parameters on the quality of welding seam. Stokes[6] studied the change of viscosity of liquid plastic during the steady melting process of rotary welding. Phadke [7] introduced the rotary welding technology of connecting spherical or cylindrical plastic parts by friction welding.

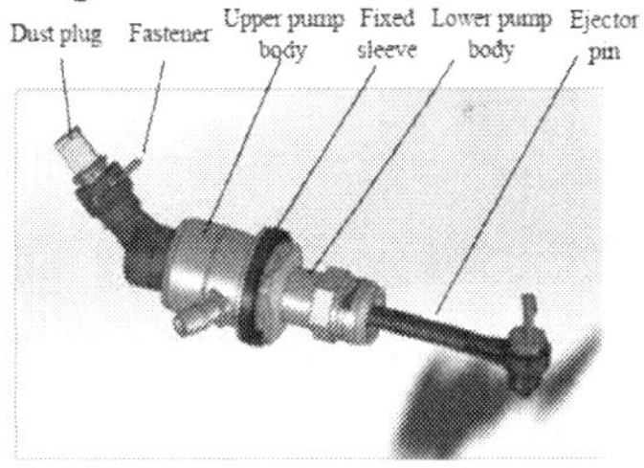

Figure 1a. Structure of a clutch master cylinder

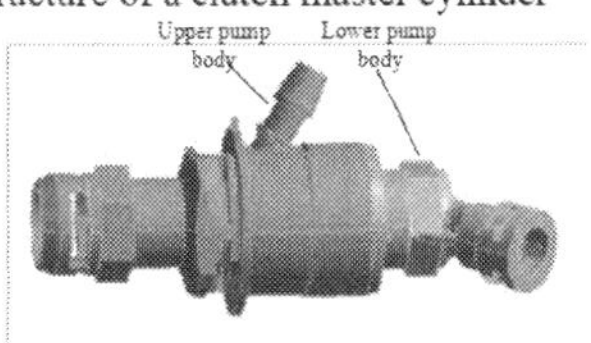

Figure 1b. Clutch pump body

Figure 1. Clutch total pump structure and clutch pump body

So far, no relevant report on the design of a clutch pump body (Fig. 1) mentioned in this paper has been seen. Although the above research has certain reference value, it is not enough to guide the structural design of a clutch

[a] Corresponding author: shengqqqqjing@163.com

pump body using friction welding. Therefore, this paper puts forward the structural design scheme of the clutch pump body based on the friction welding technology. Through the strength analysis of the clutch pump body and its weld, as well as the test verification, the correctness of the structural design of a clutch pump body is clarified.

2 Structure Design of Clutch Pump Body

The design requirements of the clutch pump body are: it can bear 14 MPa pressure under normal temperature; it can work 7×10^4 times under 100 °C high temperature and 3.5 Mpa pressure; it can work 7×10^5 times under normal temperature and 3.5 Mpa pressure.

2.1 Mechanical model of the working state of a clutch pump body

In the elastic state, the inner wall of the pump body bears radial stress, axial stress, and tangential stress. For the thick wall pump body with internal pressure, the most dangerous point is the inner wall. The three principal stresses of dangerous points are:

$$\sigma_1=\sigma_t=\frac{K^2+1}{K^2-1}p \tag{1}$$

$$\sigma_2=\sigma_a=\frac{1}{K^2-1}p \tag{2}$$

$$\sigma_3=\sigma_r=-p \tag{3}$$

In the formula, $K=D/d$, σr is the radial stress, σt is the tangential stress, σa is the axial stress, D is the outer diameter of the pump body, d is the inner diameter of the pump body, P is the internal pressure of the pump body. When the pump body is in normal operation, the cavity is filled with hydraulic oil, and the oil pressure is 3.5 MPa. Considering the design requirements of a clutch pump, set the oil pressure to 14 MPa (safety factor is 4). According to the first and third strength theory, the mechanical model of the clutch pump is:

$$\sigma_1=\frac{K^2+1}{K^2-1}p\le[\sigma] \tag{4}$$

$$\sigma_1-\sigma_3=\frac{2K^2}{K^2-1}p\le[\sigma] \tag{5}$$

In the formula, $[\sigma]$ is the allowable stress of the material.

2.2 Material selection of clutch pump body

PAMXD6 has good mechanical properties (Tab. 1), heat resistance, oil resistance, weak acid resistance, alkali resistance, good shock absorption, and noise reduction, therefore it is widely used in automobiles, electrical equipment, and mechanical parts. Based on the technical requirements of the clutch pump body and consideration of production cost, PAMXD6 is selected as the material for the manufacturing clutch pump body.

Table 1. Performance Parameters of PAMXD6 Material

Parameter	Value
Density	1.70 g/cm^3
Tensile strength	255 MPa
Elongation at break	1.9 %
Modulus of elasticity	20000 MPa
Rockwell hardness	100 M

2.3 Structure design of welding surface of clutch pump body (upper and lower pump bodies) considering the friction welding process

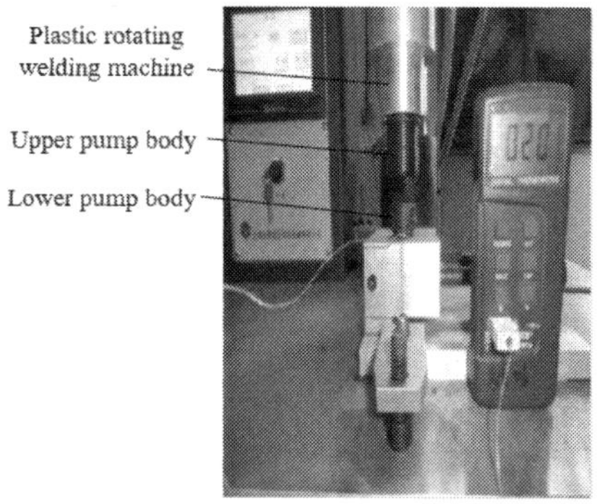

Figure 2. Welding Equipment of the Pump Body of the Plastic-Made Clutch

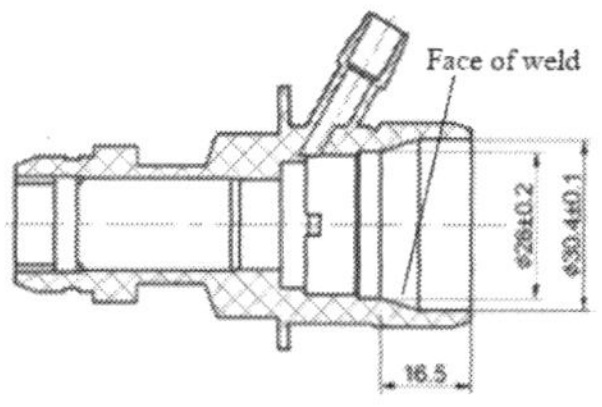

Figure 3a. Cylinder welded upper pump body

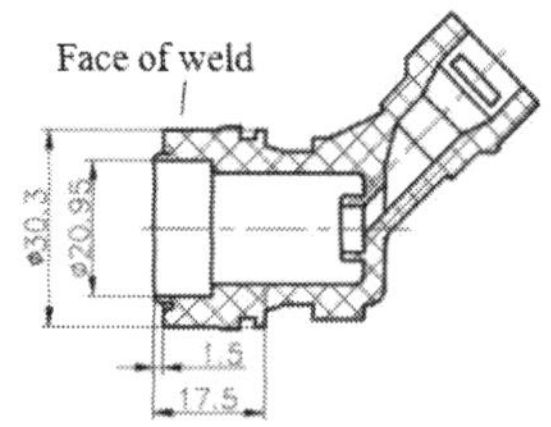

Figure 3b. Cylinder welded lower pump body
Figure 3. Plastic clutch pump body with cylindrical welding style

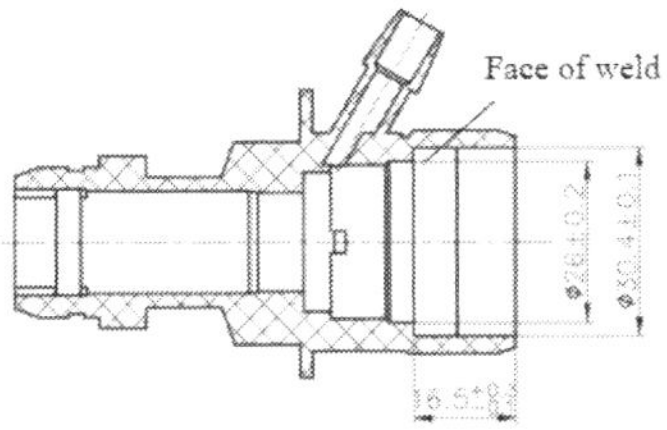

Figure 4a. End welded upper pump body

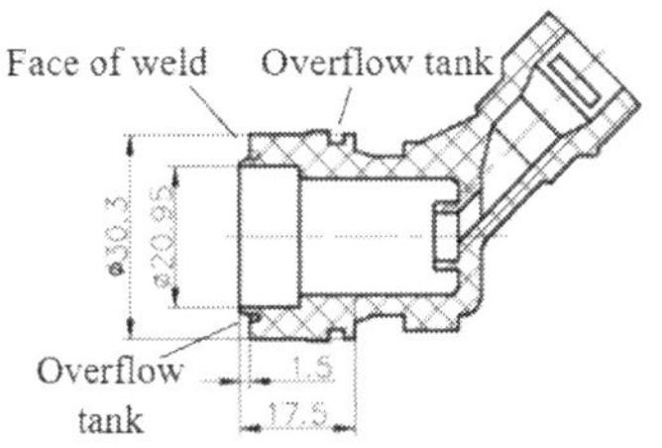

Figure 4b.End welded lower pump body
Figure 4. Plastic clutch pump body with end face welding style

The process of rotary friction is: contact, press and rotate the welding surfaces of the two plastic parts, resulting in the friction heat to melt the contact surfaces; after the rotation stops, the pressure is maintained until the melted materials solidify, as shown in Fig. 2. When designing the welding surface, the welding surface should be simple in structure and compact in space. According to the functional requirements of the clutch pump, material characteristics and mechanical model of the clutch pump body at work, combined with the characteristics of rotary friction welding, the structural types of the upper and lower pump bodies of cylinder welding type (Fig. 3) and end welding type (Fig. 4) are proposed. Considering the blown off of plastic debris during welding, the overflow tank is set up on the lower pump body.

Obviously, the lower pump body structure of the two welding methods is the same, and the upper pump body is different. When the cylinder welding type is adopted, the main shaft drives the upper pump body. Due to the elastic deformation of the material, the inner conical surface of the upper pump body contacts with the welded cylindrical surface of the lower pump body; when the end welding type is adopted, the inner end surface of the upper pump body contacts with the outer end surface of the lower pump body. When the upper pump body rotates with the main shaft of the welding machine, the two welding surfaces of the upper and lower pump body begin to be welded. The melted plastic fills the welding surfaces of the upper and lower pump body and finally connects them into a pump body assembly.

It is not difficult to see that when cylinder welding is adopted, the contact stress between the inner cone welding surface of the upper pump body and the cylinder welding surface of the lower pump body changes in gradient. With the melting of the plastic on the welding surface, the contact stress of the welding surface gradually decreases. Even if there is certain stress on the welding surface, the temperature generated will not only prevent the contact surface from melting [8], but also cause the debris to mix into the molten plastic during friction. On the way of end face welding, when the welding surface of the inner end face of the upper pump body contacts with the welding surface of the outer end face of the lower pump body, the pressure stress is uniform, and the friction heat generated per unit time is the same, which ensures the stability of the plastic melting process.

Five groups of samples were welded according to the cylinder welding method and the end face welding method, so as to verify the above analysis. In order to check the welding quality, the samples were cut. It found that the welding seam of the sample after cylinder welding is uneven to different degrees, and the welding seam is obviously mixed with debris (see Fig. 5a); However, the welding seam of the end face weldment is even and flat (see Fig. 5b), and the debris in the weld joint is not obvious. In order to further verify the welding quality of the two welding methods, 10 samples of the above two methods were tested for the weld strength. The results showed that the average pressure of the clutch pump body manufactured by the cylinder welding method was 4.4mpa, and the average pressure of the clutch pump body manufactured by the end face welding method was 9.6MPa.

Figure 5a. The welding seam of cylindrical welding

Figure 5b. The welding seam of end face welding
Figure 5. Weld condition

Through the comparative analysis and test results, it is determined that the structure of the welding surface of the upper and lower pump body is the end face welding type.

3 Strength Analysis of Clutch Pump Body

After completing the preliminary design of the clutch pump body, it is necessary to conduct strength analysis to check whether the clutch pump body meets the design requirements.

3.1 Static analysis of clutch pump body

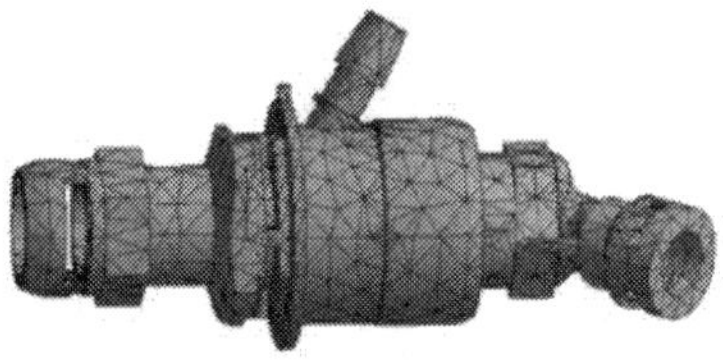

Figure 6. Simplified Model of Clutch Pump Body

The key index in the design requirements of a clutch master cylinder: it can bear 14MPa pressure. Aiming at this technical requirement, the corresponding pump body is required to meet. Therefore, it is necessary to analyze the pump body, especially pay attention to the stress condition of the welding seams. Due to the complex structures of the upper and lower pump bodies (many chamfers and fillets), the solid model of the pump body is simplified according to Saint Venant's theorem, and the simplified model of the clutch pump body is shown in Fig. 6.

3.1.1 Meshing

The simplified model of the pump body shown in Fig. 6 is imported into ANSYS Workbench, the material is PAMXD6, the elastic modulus and density are set according to the values in Tab. 1, the Poisson ratio is 0.3, and the tetrahedron mesh in the free mesh is used. The number of units is 12251, and the number of nodes is 22454.

3.1.2 Exert constraints and boundary conditions

In accordance with the limit condition of the automobile plastic clutch master pump, the force analysis of the pump body is carried out. On the basis of the design requirements, a uniform load of 14MPa (the normal working pressure is 3.5Mpa) is applied on the cylindrical surface of the upper and lower pump bodies, as shown in Fig. 7. In line with the actual working condition of the clutch pump, the fixed constraint is applied to the flange surface of the upper pump body and the cylinder surface of the lower pump body (cylinder of the welding surface), as shown in Figure 7.

Figure 7. Distribution of fixed constraint of clutch pump body

3.1.3 Stress and strain analysis of clutch pump body

Fig. 8a, Fig. 8b and Fig. 8C are stress, strain and deformation diagrams of the clutch pump body respectively.

Figure 8a. Stress diagram of the clutch pump body

Figure 8b. Strain diagram of the clutch pump body
Figure 8. Finite Element Analysis of Clutch Pump Body

It can be seen from Fig. 8a, Fig. 8b and Fig. 9C that the stress, strain, and deformation near the welding surface are relatively small. The stress is concentrated near the port of the upper pump body. The maximum stress and strain are 199.25MPa and 0.01237mm respectively. Because the maximum stress of the stress pump body is 199.25MPa, which is less than the allowable stress (255MPa), the pump body strength meets the design requirements.

3.2 Static analysis of the welding seam of the clutch pump body

The upper pump body was removed and the lower pump body is taken as the research object to carry out the static analysis of the welding seam. The stress and strain analysis results of the pump body weld are shown in Fig. 10a and fig. 10b.

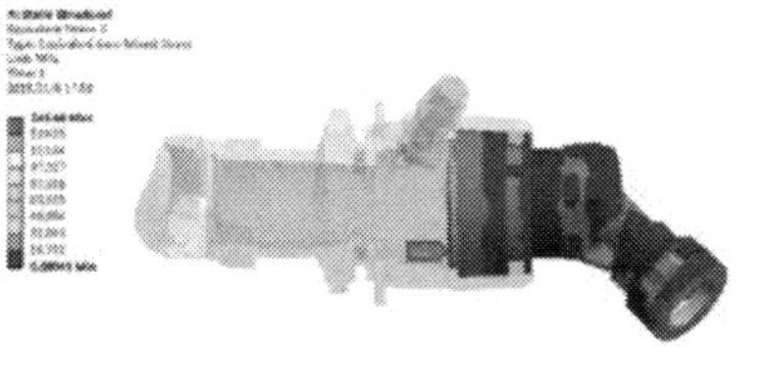

Figure 9a. Stress diagram of pump body weld

Figure 9b. Strain diagram of pump body weld
Figure 9. Stress and strain at weld of pump body

It can be seen from Fig. 9a and Fig. 9b that the maximum stress and strain occur at the weld joint of upper and lower pump bodies. The maximum stress and strain at the welding seam are 145.66MPa and 0.007676mm. Because the stress at the weld (145.66MPa) is less than the allowable stress of the material (204 MPa) (the allowable stress at the weld is 0.8 times of the allowable stress of the material), the weld strength meets the design requirements.

4 Experimental Parts

After the upper and lower pump bodies were processed by injection molding and drying (120°C, 12h), the welding parameters of the test samples are designed by the methods mentioned in literature [8], and the upper and lower pump bodies are welded in the welding equipments (EVA EGP-3p, see Fig. 2). The parameters of friction rotary welding designed in this paper are as follows: rotating speed 3000 r/ min, welding pressure 5MPa and number of turns 16r. In the light of the design requirements of the clutch pump body, pressure test, high-temperature durability test, and normal temperature durability test are carried out.

4.1 Clutch pump body pressure test

The pressure test equipment (Fig. 10) is used to test the maximum pressure on the clutch pump body.

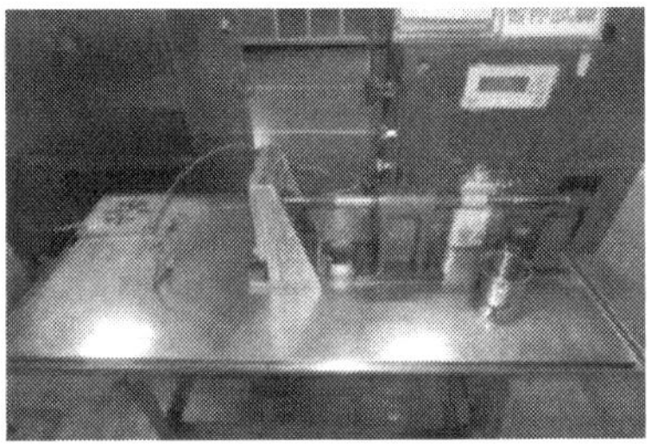

Figure 10a. Clutch pump pressure test equipment

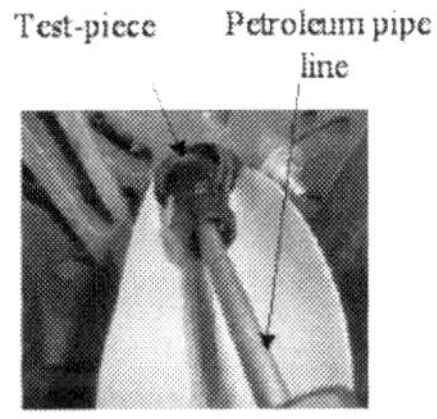

Figure 10b. Specimen test
Figure 10. High-pressure testing machine and test piece

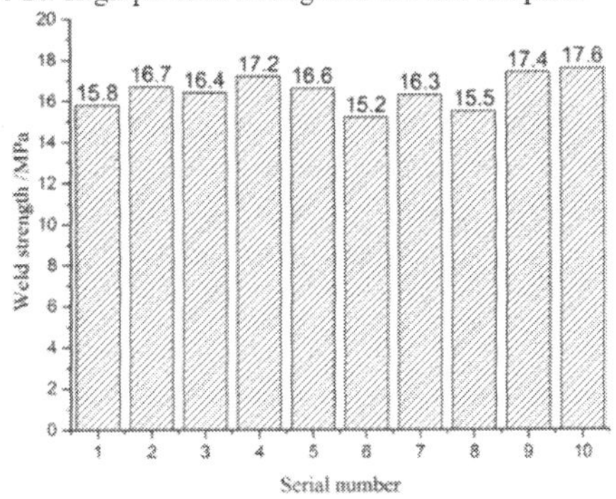

Figure 11.Welding strength of 10 groups of specimens

What can be seen from Fig. 11 is that according to the pump body structure and specified welding process proposed in this paper, the minimum pressure of 10 groups of plastic clutch pump bodies (including weld) after friction welding is 15.2MPa, the maximum pressure is 17.6MPa, exceeding the design requirements of 14MPa. The test results show that the end welded upper and lower pump body structure can meet the requirements of weld strength after welding.

4.2 Durability tests

5 groups of test pieces are selected (the number of pieces that can be tested at the same time by the equipment is 6) and put into the high-temperature endurance test equipment with the set temperature of 100°C (Fig. 12), and the pressure is built with the standard circulation of 3.5MPa in each cylinder. After the test of 5 groups of test pieces, no oil leakage of the clutch pump body (including weld) is found when the number of cycles reaches 72870.

Figure 12a. High-temperature endurance test equipment

Figure 12b.Test condition of a test piece
Figure 12. High-temperature endurance test bed and test pieces

Similarly, five groups of test pieces are selected and put into the normal temperature (25°C) endurance test machine (Fig. 13) to build the pressure, with the standard circulation pressure of 3.5MPa for each pump body. Through the test, no oil leakage of the clutch pump body (including weld) is found after the number of cycles of 5 groups of test pieces reaches 734800.

Figure 13. Room temperature durability experimental equipment

By means of the high-temperature durability test of 100 ℃ and 3.5Mpa pressure, and the normal temperature durability test of 25 ℃ and 3.5MPa pressure, the average life of the test piece is more than 7×104 times and 7×105 times respectively. The test results show that the structural design of the clutch pump body meets the design requirements.

5 Conclusions

Through the structural analysis and design of the master cylinder of an automobile plastic clutch, the conclusions are as follows:

1. The structure of the upper and lower clutch pump bodies with end face welding is put forward. By comparative analysis with the structural types of the upper and lower pump bodies of the cylindrical welded type in terms of welding quality, the structural type of the upper and lower pump bodies of the end face welded clutch is considered reasonable. The fact that the clutch pump body has achieved mass production fully illustrates this point.

2. By the finite element analysis, the strength of the clutch pump body and welding seam meets the technical requirements. Through the establishment of the finite element models of the pump body, the static analysis of the clutch pump body and the welding seam is carried out. The maximum stress of the pump body and the welding seam is 199.25MPa and 145.66MPa respectively, which meets the strength requirements. In addition, the stress of the upper pump body, the lower pump body, the dangerous section, and the weld under the working condition is also studied, and the results meet the first and third strength theory.

3. After testing, the pressure and durability of the clutch pump body meet the design requirements. The pressure of the trial clutch pump body (including weld) exceeds the design requirements of 14MPa, and the number of durability test cycles at high temperature and the normal temperature reaches 72870 and 734802 respectively, which meet the design requirements.

4. The relationship between weld formation mechanism, weld shape, and weld strength needs to be studied. The next research work will focus on: the process of friction rotary welding, the process of weld formation and its influencing factors; the influence of weld size and shape on welding strength.

Acknowledgments

The authors would like to acknowledge the financial support from the Xiamen Science and Technology Project (grant No. 3502Z20179023), the Fujian Science and Technology Flatform Project (grant No. 2016BJ001), the Science& Technology Innovation Project of Fujian Province (grant No. 2016H2003) ,and the Educational Research Project for Young Teachers in Fujian Province(grant No. JT180449).

References

1. A. H. Singh , S. Harmeet , P. Aashish, OEM's approach to the design and evaluation of plastic clamps, SAE Int. J. of Passenger Cars - Mechanical Systems , **10**(1), 141-149 (2017)
2. D. J. Liang, Q. G. Deng, X. M. Wan, Optimization design of automobile plastic parts based on GRNN neural network, Plastics, **3**, 98-102 (2018)
3. W. X. Zhang, H. J. Chang, Y. L. Dou, Polyurethane slush powders for instrument panel of automobiles: composition, Processing and Rheological Properties, Materials Reports, **32**(24),185-190 (2018).
4. W.Hou, Y. Qian, Z. X. Su. Application of all-plastic structure of new energy vehicles in the improvement of crash safety, China Plastics Industry, **45**(7),140-143 (2017).
5. P. Tappe, H. Potente, New results on the spin welding of plastics, Polymer Engineering &Science, **29**(23), 1665-1660 (1989)
6. V. K. Stokes, A. J. Poslinski. Effects of variable viscosity on the steady melting of thermoplastics during spin welding, Polymer Engineering &Science **35**(5), 441-459 (1995)
7. N. L.Phadke, Spin welding, Popular Plastics & Packaging **60**(1), 30-32 (2015)
8. J. Sheng, R.Y. Xie, A design method of processing parameters of rotary friction welding for a plastic clutch pump ,China Mechanical Engineering, **30**(14), 1742-1747 (2019)

Real-time Reliability Evaluation Based on Stochastic Degradation Process and Sequential Joint Distribution Estimation

QingZhen Gao[1,a], LiHu Teng[2], Zhuo Jiang[3] and Hua Zhang[4]

[1]Anhui Polytechnic University, School of Mechanical and Automotive Engineering, 241000 Wu Hu, China
[2]Cheng Du Bureau of The Air Force, The Second Military Delegate Room, 550000 Gui Yang, China
[3]Air Force Armament Department, Audit Division, 100000 Bei Jing, China
[4]AVIC Guizhou Liyang Aviation Power Co., Ltd, 550000 Gui Yang, China

Abstract. According to the different performance degradation paths, reliability curves and parameters of each product in service phase, the reliability and remaining life of components at current moment can be evaluated in real-time reliability assessment by integrating real-time status, historical information and service time of components. The real-time reliability evaluation is valued because it is more personalized, precise, real-time and lean than traditional methods. Furthermore, the results from real-time reliability assessment can also represent the health status of component at current time. In the current research, real-time reliability assessment mainly relies on regression analysis and time series analysis, but these two methods are mainly used to describe the component degradation process, and cannot reflect the influence of external random environment on the component state change. At the same time, due to these limitations of time, economy and test conditions, it is also worth studying how to obtain more accurate and practical reliability distribution and determine a detection interval under the condition of less data. Therefore, based on the analysis of real-time reliability evaluation principle, a more appropriate real-time reliability evaluation method and the detection interval decision model are proposed by means of random degradation process, parameter sequence test joint distribution estimation and failure risk. The result from this method is a quantized value, which is convenient for the direct application of the follow-up maintenance decision research. Therefore, the research in this paper has extensive reference value and practical application prospect.

1 Introduction

Real-time reliability assessment[1] refers to the real-time evaluation for reliability index based on real-time detection information from a specific component or system. The principle of real-time reliability evaluation is shown in Figure 1.

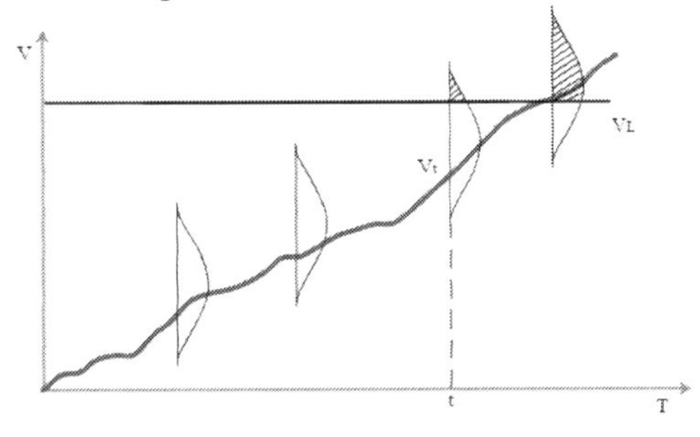

Figure 1. Schematic diagram of real-time reliability assessment of actual performance degradation process.

At any time t, the value of performance variable V_t meets a certain distribution. In figure 1, the value of hypothetical energy variable in the figure is normal distribution. And the fixed failure threshold V_L is used as the failure distribution $F(t, V_L)=P(v(t)>V_L)$. That is the shaded part of the figure, so the real-time reliability is $R(t)=1-F(t, V_L)$.

Suppose that there is a kind of product after the experiment of large sample size, the performance degradation process and its distribution obtained by statistical analysis are shown in Figure 2.

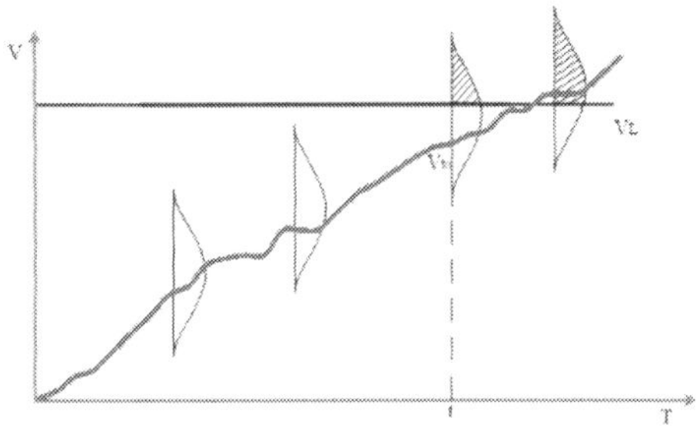

Figure 2. Performance degradation process and its prior distribution based on statistical analysis.

[a] Corresponding author: gaoqingzhen68@tom.com

The shaded part in figure 1 can be considered as the average reliability of similar products at that time, and the average reliability is the result of statistical analysis. The performance degradation process of a real product in this category is showed in figure 1. Comparing figure 1 and figure 2, after the same time t, the statistical average degradation quantity is V_{t_0}, while the actual degradation quantity is V_t, $V_t \neq V_{t_0}$. Therefore, the distribution of performance parameters at this time in figure 1 should be different from that in figure 2. Comparing figure 1 and figure 2, since the actual degradation amount of the product is smaller than the average degradation amount. So the shaded part in figure 1 is smaller than the shaded part in figure 2. Which means the failure probability is small, that is the reliability is higher than the average reliability. The distribution of performance parameters in figure 1 includes both historical experimental data analysis and field measurement data. So it can better represent the actual situation of the product. However, the distribution of figure 1 is unknown.

The primary task of real-time reliability assessment is to obtain the prior distribution in figure 2 by statistical analysis of the historical information of similar products, and then to obtain the distribution of performance parameters in figure 1 which integrates real data with the state data measured in the field, so as to obtain the product real-time reliability.

The research methods of real-time reliability assessment mainly include performance degradation analysis method and real-time reliability index method. Among them, the performance degradation analysis method believes that the reliability decline of components or systems is caused by internal performance degradation. In other words, the product is considered to have failed when the performance degradation parameter value of a product reaches the pre-set threshold. Therefore, real-time reliability can be evaluated by establishing the corresponding performance degradation model.

Zhe Sun and linhai Zhao[2] established an adaptive cloud model of failure efficiency between ambient data and component failure efficiency based on cloud theory. Zhixue Tan et al[3] proposed a specific TLML algorithm to specified the loss function and alternative training scheme of component models for harnessing the information from sensor readings and empirical rules to serve the modeling. In literature[4], an adaptive Kriging-based reliability analysis method was proposed to estimate the prior failure probability and the conditional probability about information. Its main contribution was a decomposition of rare joint failure event into two events both with relatively high probabilities. Zhiliang et al[5] established a reliability model of marine power station under different working conditions and a reliability model database. Cai Baoping and Liu Yonghong[6] proposed a novel real-time reliability evaluation methodology by combining root cause diagnosis phase based on Bayesian networks (BNs) and reliability evaluation phase based on dynamic BNs (DBNs). Yu Hongping and Tang Mao[7] studied reliability assessment of multi-component systems under competing degradation and random shocks. Jinbo Huang et al[8] presented a degradation model with a random failure threshold in reliability assessment by Bayesian approach.

In summary, regression analysis and time series analysis are mainly used to describe the degradation process for a system or component in current real-time reliability studies. However, these methods cannot reflect the influence of external random environment on component state change. Therefore, the methods of real-time reliability evaluation and detection interval decision are proposed in this paper based on random process, sequential joint distribution and failure risk theory.

2 The Analysis of evaluation methods

According to the principle of real-time reliability assessment, the complete reliability assessment process includes that firstly the appropriate degradation process is selected to represent the actual degradation process of components, then the prior distribution of unknown parameters in gradation process is determined and the posterior distribution is obtained by bayesian method, finally the reliability assessment model is derived based on failure threshold value. The specific steps are as follows:

1.The appropriate distribution $G(x,\theta(t))$ is selected to represent the distribution of performance degradation quantity of component at time, where $\theta(t)$ is the unknown random parameter, and the probability density function $G(x,\theta(t))$ is $g(x,\theta(t))$.

2.Based on the historical information of similar components, the prior density function $\pi(\theta(t_i))$ of $\theta(t)$ at each time point t_i is determined by statistical analysis.

3. According to the performance degradation quantity V_i actually measured at time t_i of the component, the unknown parameter posterior density function $\pi(\theta(t_i)|V_i)$ is obtained by bayesian method, then the probability density function $g(x,\theta(t_i)|V_i)$ of the performance degradation quantity is obtained.

4. According to $g(x,\theta(t_i)|V_i)$ and the set failure threshold l, the component reliability $R(t_i)$ at time t_i is obtained, that is the real-time reliability evaluation model for the component at time t_i.

2.1 The model of stochastic degradation process

The component performance degradation and failure is a slow development process. Environmental stress causes component to produce wear and damage continually. Wear and damage accumulation process is the component degradation process. A lot of components are subjected to continuous impact damage in actual work environment, such as vibration damage, overload damage, etc. The occurrence of impact damage may be time continuous or discrete random. When the cumulative damage reaches a certain level, it leads to component failure.

For the stochastic process model of continuous damage accumulation, Wiener-einstein degradation failure model and Gamma degradation failure model[9] are the main models. Since most actual degradation processes are monotonous and random, so their increments are nonnegative and independent. It is very appropriate to use Gamma processes to describe their degradation processes.

It is assumed that the performance parameter variable $X(t)$ obeys Gamma distribution at any time t and its distribution density function is:

$$f_{X(t)}(x) = Ga(x|v(t),u) = \frac{u^{v(t)}x^{v(t)-1}\mathrm{e}^{-ux}}{\Gamma(v(t))} I_{(0,\infty)}(x) \quad (1)$$

Where, $I(\cdot)$ is the function which is constant 1 but not constant 0 in the domain, $\Gamma(\cdot)$ is the Gamma function, $v(t)$ and u are the shape parameter and scale parameter of the Gamma distribution function $Ga(\cdot)$ respectively, $v(t)$ is the continuous and non-subtractive real function, $v(0) \equiv 0$, and u is the real number greater than zero. Assuming that the Gamma process is a stationary process and it can be set $v(t)$ as a linear function $v(t) = at$, where a is a constant. The expectation of performance parameter variable $X(t)$ is:

$$E(X(t)) = \int_0^{\infty} x f_{X(t)}(x)dx = \frac{v(t)}{u} = \frac{at}{u} \quad (2)$$

The variance is:

$$E\left((X(t)-E(X(t)))^2\right) = \frac{at}{u^2} \quad (3)$$

Let $E(X(t)) = st$, $E\left((X(t)-E(X(t)))^2\right) = m^2 t$, and $m = vs$, v is a constant, then the $X(t)$ distribution function expression (3) can be transformed into $f_{X(t)}(x) = Ga(x|t/v^2, 1/sv^2)$. At this point, the unknown random parameter of distribution function is s.

2.2 The sequential distribution estimation

For a specific function, the form of prior distribution can be obtained by the conjugate distribution method, and the inference process from prior distribution to posterior distribution can be simplified. Firstly, the conjugate prior distribution form of parameter s is assumed, and then the parameter is determined according to historical information, that is the prior distribution of parameter s.

Let the prior distribution of parameter s be the inverse Gamma distribution, and the probability density function is:

$$\pi(s) = Ig(s|\delta,\eta) = \frac{\eta^{\delta}}{\Gamma(\delta)}\left(\frac{1}{s}\right)^{\delta+1}\mathrm{e}^{-\frac{\eta}{s}} I_{(0,\infty)}(x) \quad (4)$$

Where: δ and η are constant respectively, the general determination method is to use historical information to fit.

Assuming the actual quantity of degradation is detected at time t is y, then the conjugate posterior probability density function of s is:

$$\pi(s|y) = \frac{f(y|s)\pi(s)}{\int_0^{+\infty} f(y|s)\pi(s)ds}$$

$$= \frac{\left(\frac{y}{v^2}+\eta\right)^{(\frac{t}{v^2}+\delta)}}{\Gamma(\frac{t}{v^2}+\delta)}\left(\frac{1}{s}\right)^{\frac{t}{v^2}+\delta+1}\mathrm{e}^{-\frac{\frac{y}{v^2}+\eta}{s}} I_{(0,\infty)}(x)$$

$$= Ig\left(s\left|\frac{t}{v^2}+\delta, \frac{y}{v^2}+\eta\right.\right) \quad (5)$$

It is assumed that there are j actual test data y_1, L, y_j at the existing time t_1, L, t_j, and the posterior joint distribution of parameter s is:

$$\pi(s|y_1,\mathrm{L},y_j) = Ig\left(s\left|\delta+\sum_{i=1}^{j}\frac{t_i-t_{i-1}}{v^2}, \eta+\sum_{i=1}^{j}\frac{y_i-y_{i-1}}{v^2}\right.\right)$$

$$= Ig\left(s\left|\delta+\frac{t_j}{v^2}, \eta+\frac{y_j}{v^2}\right.\right) \quad (6)$$

The above formula indicates that the calculation of posterior probability density function requires only the last detection result, and the estimated value of posterior expectation for s is:

$$\hat{s} = \frac{\eta + y_j/v^2}{\delta + t_j/v^2 - 1} \quad (7)$$

The formula (7) is substituted into equation (3) to obtain the posterior distribution density function of performance parameter variables.

$$f_{X(t)}(x) = \frac{(1/\hat{s}v^2)^{t/v^2} x^{t/v^2-1}\mathrm{e}^{-x/\hat{s}v^2}}{\Gamma(t/v^2)} I_{(0,\infty)}(x) \quad (8)$$

2.3 The real-time reliability assessment

Suppose the degradation failure threshold of component is l, then the detection result y_j at t_j moment can be calculated according to the reliability definition.

$$R(t_j) = P\{x(t_j) < l\} = \int_0^l f_{X(t_j)}(x)dx$$

$$= \int_0^l \frac{(1/\hat{s}v^2)^{t_j/v^2} x^{t_j/v^2-1}\mathrm{e}^{-x/\hat{s}v^2}}{\Gamma(t_j/v^2)} dx \quad (9)$$

The formula (9) is the real-time reliability evaluation model for component with the detection result y_j at time t_j.

2.4 Determination of detection interval based on failure risk

Based on the above real-time reliability research, the failure conditional probability is $F(t+\Delta t|t)$ in the next time period Δt. If the component is detected and found to be working normally from time t. Suppose the life of component is T, then according to conditional probability, the risk of component failure can be expressed as:

$$F(t+\Delta t|t)=\frac{P(t<T)-P((t+\Delta t)<T)}{P(t<T)} \quad (10)$$

According to the reliability definition $P(t<T)=R(t)$, formula (10) can be evolved into formula (11).

$$F(t+\Delta t|t)=\frac{R(t)-R(t+\Delta t)}{R(t)} \quad (11)$$

Assume the fault risk threshold is constant D, that is the fault risk is not greater than D. In general, as long as an appropriate Δt is found, so that the fault risk within the range of Δt is equal to the fault risk threshold D. Then this Δt is the optimal detection interval. The Δt meets the requirement of the longest detection interval under fault risk constraint, then formula (11) is converted to formula (12).

$$\frac{R(t)-R(t+\Delta t)}{R(t)}=D \quad (12)$$

Since the increment of Gamma process is independent and non-negative, it is very suitable for describing the random process of monotone degradation. Hypothesis: (1) At any measurement time t, the performance parameter variable $X(t)$ follows the Gamma distribution. (2) The traditional reliability function $R_c(t)$ of component obeys Weibull distribution, that is $R_c(t)=\exp\left(-(t/\lambda)^{\beta}\right)$.

Therefore, the first detection interval t_1 of the component is:

$$t_1=\lambda\left(-\ln(1-D)\right)^{\frac{1}{\beta}} \quad (13)$$

According to the real-time reliability function $R(t)=\exp\left(-(t/\hat{\lambda})^{\hat{\beta}}\right)$, the detection interval Δt_i determined at time t_i after the first state detection is :

$$\Delta t_i=\hat{\lambda}\left((\frac{t_i}{\hat{\lambda}})^{\hat{\beta}}-\ln(1-D)\right)^{\frac{1}{\hat{\beta}}}-t_i \quad (14)$$

3 An example analysis

The table 1 is the data record of percentage change with time about working current at the temperature of 90°C. The working current data is from an engine pressure sensor that is a high reliability and long life component. A total of 15 samples were tested, and the data were tested every 500 hours until 4000 hours. It is assumed that the failure threshold of the sensor is 10.

Table 1. A record of percentage change with time about working current of a sensor at 90℃.

	0	500	1000	1500	2000	2500	3000	3500	4000
1	0	0.95	2.75	4.36	5.50	6.75	8.03	9.52	10.97
2	0	1.24	2.33	3.77	5.02	6.09	7.19	8.45	9.31
3	0	1.19	2.01	2.99	3.96	4.47	5.29	6.03	6.91
4	0	0.64	1.97	2.97	3.82	4.53	4.50	5.64	6.17
5	0	0.63	1.77	2.60	3.40	4.66	5.64	6.35	7.62
6	0	1.41	2.91	3.84	5.37	6.73	8.65	9.99	11.04
7	0	0.95	1.49	2.42	2.97	4.51	4.87	5.58	7.19
8	0	1.09	1.79	2.43	3.05	3.77	4.78	5.48	6.27
9	0	0.95	1.98	3.45	4.14	4.94	5.87	6.88	7.81
10	0	1.51	3.03	4.52	6.28	7.82	8.95	10.47	12.25
11	0	1.02	1.99	2.86	4.04	4.83	5.68	6.56	7.46
12	0	0.82	1.76	3.63	4.46	5.25	5.99	7.02	7.91
13	0	0.76	1.87	2.97	3.95	5.50	6.54	7.42	8.93
14	0	0.72	1.37	2.58	3.42	4.51	5.44	6.17	6.70
15	0	0.85	1.56	2.30	3.46	4.13	4.65	5.88	6.66

From the 135 data in table 1, the trait of performance data degradation with time can be seen: The initial quantities are all 0; Monotonic growth; Increments are independent. Thus the degradation of sensor performance can be described by the Gamma process.

After the Gamma process model was established, the prior distribution of parameter s must be determined. According to equation (4), the prior distribution s is determined by δ and η. δ and η can be estimated from the data in table 1. After analysis, the δ=1.4, η=6 are taken for inverse Gamma distribution as the prior distribution of parameter s. The reliability evaluation model for working current of the sensor at 90℃ can be obtained in formula (15) from equations (7) and (9) by taking v=1.

$$R(t_j)=\int_0^l \frac{(\frac{1.4+t_j-1}{6+y_j})^{t_j} x^{t_j-1} e^{-x\frac{1.4+t_j-1}{6+y_j}}}{\Gamma(t_j)}dx \quad (15)$$

It can be seen that the reliability assessment model not only considers the time factor, but also considers the real-time test data.

The data at 2000 hours from the fifth sample is used for experimental calculation. By substituting l=10, t_{200}=2000, y_{2000}=3.40 into the equation, the real-time reliability at that time $R(t_{2000})$=0.9980 can be obtained.

The traditional reliability assessment[10] was finished based on the data in table 1. At a given time t, the traditional reliability is $R'(t)=\exp(-(t/5818.6)^{5.62})$. If the date at 2000 hours is substituted in this equation, the average reliability of these 15 samples at 2000 hours is $R'(2000)$=0.9974.

Comparing the real-time reliability of the fifth sample with the average reliability from all 15 samples, it can be easily seen that the degradation rate of the fifth sample is slower than the average speed. So the real-time reliability of the fifth sample is higher than the average reliability at 2000 hours. In other words, the fifth sample was healthier than average at 2000 hours.

Assuming that the traditional reliability curve of the sensor follows weibull distribution, where λ =5676.5, β=5.85. Then seven data points are taken from it to form an initial data sequence. E_0 ={(1000,0.9999), (1500,0.9995), (2000,0.9975), (2500,0.9913), (3000,0.9761), (3500,0.9441), (4000, 0.8854)}. Suppose that the failure risk constraint D=0.1 during the detection interval. The λ, β and D are substituted into equation (13) to obtain the first detection interval t_1 (Per unit time).

$$t_1 = 5676.5\left(-\ln(1-0.1)\right)^{\frac{1}{5.85}} \approx 3800$$

By taking v=1, the real-time reliability $R(t_1) = 0.9565$ of the sensor in the first state detection can be obtained from formula (9). And $\hat{\beta}$ =7.6572, $\hat{r}$=65.5489, $\hat{\lambda}$ =5500 can be calculated from E_0 and the real-time reliability function $R(t) = \exp\left(-(t/\hat{\lambda})^{\hat{\beta}}\right)$. The $\hat{\beta}$, $\hat{r}$, $\hat{\lambda}$ and t_1 are substituted into equation (14) to obtain detection interval Δt_1=550.0(Per unit time) after the first state detection.

4 Conclusion

Based on studying the principle of real-time reliability assessment, a more suitable method is proposed in this paper for real-time reliability assessment and detection interval prediction of components by integrating real-time status, historical information, environmental factors and service time of components.

In this method, stochastic process is used to describe the component degradation process to make it more realistic. The problem that a component reliability distribution is difficult to obtain under the condition of less data is solved by successive joint distribution estimation. The calculation methods for the first and subsequent component detection interval are established by introducing the concept of failure risk. The above research provides reference for evaluating a component reliability in real time and scientifically.

The working data of a sensor was taken as the research object. Then according to the real-time reliability evaluation method in this paper, the real-time health status of the sample was obtained and the first and subsequent detection interval were calculated. These works verified the correctness and effectiveness of the research.

The results from this real-time reliability evaluation method and detection interval decision model are quantized values, which is convenient for the direct application of the follow-up maintenance decision research. At the same time, these factors such as the availability, accessibility and cost of information collection are fully taken into account in the method, so that the method has a good practical value and application basis.

References

1. Chunhui Liang. The research on real-time reliability analysis and prediction to spindle bearing system of wind turbine[J]. Changchun Inst. Tech. (Nat. Sci. Edi.), 2017, Vol.**18**, No.3: 41-45.
2. Zhe Sun, linhai Zhao. Real-time evaluation method of JTC reliability based on cloud theory[J]. Railway journal, 2019, **18**:53-73.
3. Zhixue Tan, Shisheng Zhong, Lin Lin. Trans-layer model learning: A hierarchical modeling strategy for real-time reliability evaluation of complex systems [J]. Engineering and System Safety, 2019, **2**(4) : 382-393.
4. Zeyu Wang, Abdollah Shafieezadeh. Real-time high-fidelity reliability updating with equality information using adaptive Kriging[J]. Reliability Engineering and System Safety, 2020, **3** (1) :154-163.
5. Zhiliang, WuYubin, YaoDan Wang. Design and Study of Real-time Online Assessing System for Marine Power Station Reliability [J]. 2017 2nd International Conference on Automation, Mechanical and Electrical Engineering, 2017: 165-168.
6. Cai Baoping, Liu Yonghong. Real-time reliability evaluation methodology based on dynamic Bayesian networks: A case study of a subsea pipe ram BOP system. [J]. ISA transactions, 2015, **5** (4): 353-357.
7. Yu Hongping, Tang Mao. Reliability assessment for systems suffering competing degradation and random shocks under fuzzy environment[J]. Science progress, 2019,**10**(1):166-179.
8. Jinbo Huang, Dejing Kong, Lirong Cui. Bayesian reliability assessment and degradation modeling with calibrations and random failure threshold[J]. Journal of Shanghai Jiaotong University (Science), 2016, Vol.**21** (4), pp.478-483.
9. Wang Xiaolin, Guo Bo, Cheng Zhi. Real-time reliability evaluation for product with nonlinear drift-based Wiener process[J]. Journal of Central South University (Science and Technology), 2013(8), Vol.**44**, No.8:3203-3209.
10. Wang Haoming, Tang Chong, Wu Liping, Yang Lei, Yao Ying, LiuHong. Dynamic Reliability Assessment of Distribution Network Based on Equipment State Evaluation Model[J]. Proceedings of the CSU-EPSA, 2017 (7), Vol.**29**, No.7: 68-74.

Project support: 2019 talent fund project of Anhui Polytechnic University (2019YQQ001)

Analysis and optimization of rubidium spectrum lamp to eliminate frequency fluctuations of rubidium atomic frequency standard

Jiqing Lian[1,2,a], Jinhai Zhang[1], Fucheng Shi[1], Jun Zhang[1], Lixun Wei[1] and Kuan Wang[1]

[1]*Lanzhou Institute of Physics, 730000 Lanzhou, China*
[2]*Science and Technology on Vacuum Technology and Physics Laboratory, 730000 Lanzhou, China*

Abstract. Rubidium atomic frequency standard (RAFS) is the most widely used frequency standard in space. The light used to pump the atoms and detect the resonance signal is emitted by rubidium spectrum lamp, so the light intensity of rubidium spectrum lamp directly determines the performance of RAFS. This paper discussed on-board RAFS' output frequency fluctuations caused by rubidium spectrum lamp. The reason of frequency fluctuations from rubidium lamps was described. To obtain stable lamp light intensity, analysis and optimization of the lamp was developed. Relevant experiments were carried out to verify the optimization. The study content of this paper is beneficial to improve the performance of a single temperature controlled space RAFS physics package.

1 Introduction

Rubidium atomic frequency standard (RAFS) is the most widely used atomic clock in space, which is characterized by small size, low power consumption and good stability. RAFS has been successfully applied to the satellites for Navigation, remote sensing, reconnaissance and other applications. The accuracy and stability of RAFS is closely related to the performance of the satellite. For example, in the electronic system of a navigation technology satellite (NTS), the on-board RAFS, as the clock signal reference of the satellite, provides accurate and stable clock signals for the satellite electronic system, and its stability index directly affects the precision of the navigation system. Unfortunately, research shows that, when the satellites operate in space, especially in the first one or two years, frequency fluctuations may happen to the rubidium atomic frequency standard, which have bad effect on the accuracy and stability of the RAFS. Although the frequency fluctuations mainly decreased with the passage of time, they were still one of the main factors affecting the performance of space RAFS. Therefore, it is of great significance to reduce the frequency fluctuations of space RAFS.

2 Description and analysis of frequency fluctuations

According to reference [1][2], most of the RAFSs on GPS IIR satellites had a general performance that is among the best of the constellation. However, there were still several RAFSs with worse performance mainly caused by frequency fluctuations. There were also frequency fluctuations in BDS satellite RAFS[3]. The fluctuations were mainly manifested in two ways: 1, the fluctuations occurred periodically with the period about a few days; 2, the frequency fluctuations occurred in the form of steps or jumps, but the magnitude of the frequency fluctuations tend to decay with time. The frequency fluctuations of the RAFSs on GPS IIR satellites were mainly manifested as the second kind, which were consistent with the ground test results. And our study is also mainly focused on the frequency jumps.

Space RAFS is subject to several anomalies that can be caused by the environment or internal factors. And frequency jumps are a common type of clock anomaly which directly impacts the performance of a satellite system, such as global navigation satellite system(GNSS)[4]. For the frequency jumps, the internal factors are more important. The study[5-7] shows that frequency jumps and drift of RAFS are usually respectively accompanied by abnormal and slow light intensity changes (as shown in figure 1), and the researchers believe that the instability of rubidium spectral light intensity is one of the most important reasons for frequency jumps. A RAFS is composed of a circuit system and a physics package. Apparently, the reasons of the lamp intensity jumps include the noise of the circuit system (such as fluctuations of DC-DC's output) and the imperfect manufactured physics package. However, by monitoring the parameters of the on-board circuits, we can mainly rule out the influence of the circuit system.

[a] Corresponding author: lianjiqing@outlook.com

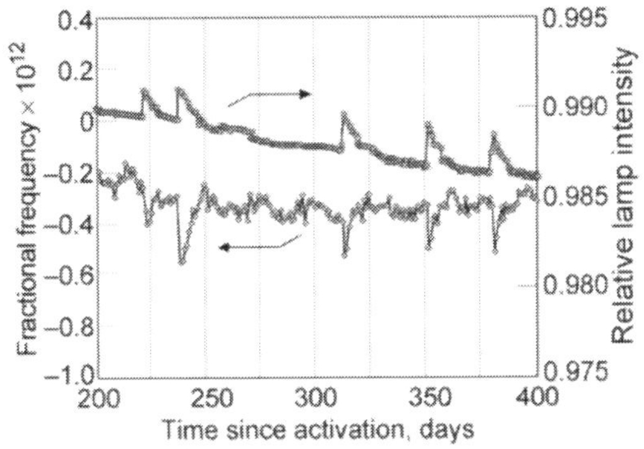

Figure 1. Jumps in the lamp intensity that coincide with jumps in the RAFS' frequency.

The physics package of a RAFS, as the block diagram shown in figure 2, contains three rubidium cells: a pure Rb-87 lamp, a discrete Rb-85 isotopic filter cell, and an Rb-87 absorption cell. The cells are usually made of alkali-free glass. When the RAFS operates, all the cells work at high temperature. In particular, the lamp's working temperature usually reaches over 100℃, which is dozens of degrees hotter than the other two cells'. The rubidium atoms in the cells will gradually decrease by permeating through the glass and chemical reaction at high temperature. In order to ensure that an on-board RAFS can operate for more than 10 years, the manufactures have to fill enough rubidium atoms to the cells[8]. The saturated vapor pressure of rubidium at a certain temperature is an invariant constant, so the excess rubidium atoms will adheres to the inner wall of the cells. There is anecdotal evidence that the frequency steps or jumps are related to excess rubidium.

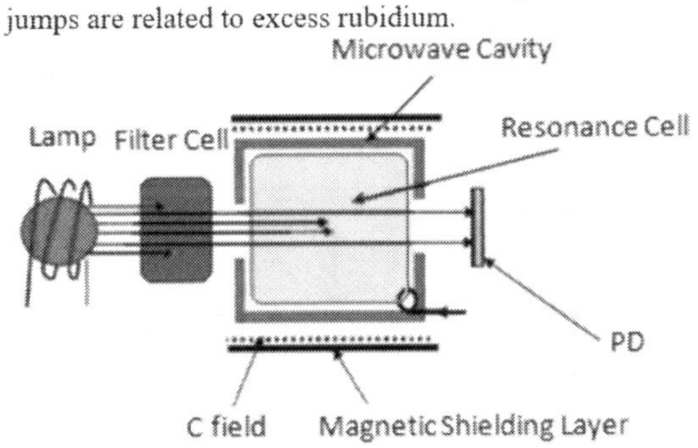

Figure 2. Schematic diagram of the physics package of RAFS.

For a RAFS, since the spectrum lamp works at a higher temperature, which leads to the faster rubidium consumption, rubidium atoms in the lamp should be more than that in the other two cells. So we believe that the excess rubidium in the lamp is the key factor of frequency jumps, and that is easily proved by testing a RAFS physics package with different lamps. A complete rubidium spectrum lamp module consists of a bulb, a exciter circuit and other structural components. Rubidium lamp exciter circuits supply about 0.5 watt of RF power at around 100 MHz to lamp bulbs. The Rb lamp is usually excited inside a series-tuned coil connected to the circuit board, which is working as a microwave inductor of the LC oscillation network. Rubidium spectrum lamp is a kind of electrodeless discharge lamp. Under the RF electric field, the discharge gas molecules (such as Ar/Kr/Xe) in the lamp bulb ionize, and then the ions transfer energy to the rubidium atoms in gas state by collisions. The rubidium atoms in excited state will decay to the ground state by emitting photons. It is clear that the RF power is directly related to the lamp brightness, and the optimization of the RF power control was reported[9], so it is not discussed in this paper.

When a rubidium spectrum lamp works in normal state, the excess rubidium atoms will be in liquid state. Usually, we make a "tail" in the end of a bulb(as shown in figure 3) to set a cold end for the accumulation of liquid rubidium. Obviously, it will take some time for the rubidium atoms in liquid state to achieve a stable distribution in the bulb. During the process of rubidium atoms' redistribution, the moving of the atoms may lead to the change of light intensity which will lead to the change of frequency. For different RAFSs, the processes of rubidium atoms' redistribution in the lamp bulbs last for different lengths of time. But overall, the distribution of rubidium in lamps tends to be more and more stable with the RAFSs' operating time. This is also consistent with the reduction of frequency jumps over operating time.

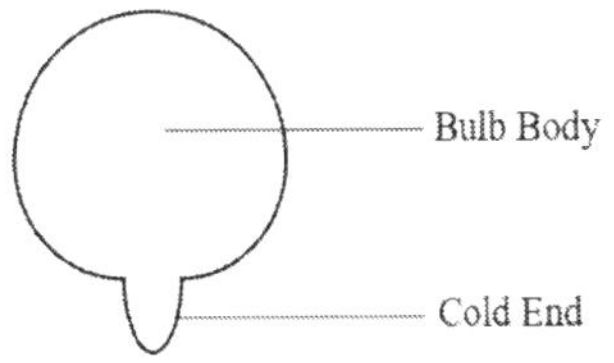

Figure 3. An typical diagram of lamp bulb.

3 Analysis and optimization of rubidium spectrum lamps

In order to abtain long-term and accurate clock signal for satellites' electric systems, frequency jumps of on-board RAFSs should be minimized or avoided. Through the analysis in the previous section, when the physics package is only considered, the frequency jumps of an on-board RAFS are mainly caused by the liquid rubidium atoms moving in the lamps to achieve the stable distribution. Rubidium atoms' stable distribution is closely related to the temperature field of lamp bulb, and the fluidity of liquid makes it difficult to hold the excess rubidium atoms in the optimal place just by optimize the manufacturing process of the lamp bulbs. Therefore, proper mechanical, electrical and thermal design of a rubidium spectrum lamp is very important, that will make the rubidium bulb obtain a temperature field more suitable and stable for liquid atoms to redistribute rapidly and smoothly.

3.1 Original design and analysis of Rb spectrum lamp

Figure 4 is the structure diagram of the rubidium spectrum lamp originally designed by Lanzhou Institute of Physics (LIP). From the bottom to the top, an aluminum baseplate, a RF excitation circuit board, an heat insulation structure, a cold end support structure, an series-tuned coil for excitation and a rubidium bulb are successively arranged. The whole lamp module is installed in an aluminium lamp room. When the RAFS operates, the RF energy is coupled to the rubidium bulb through the coil, and the temperature of the rubidium bulb is around 110℃, and there must be a temperature gradient field over the wall of the bulb because of the cold end. When the rubidium bulb reaches thermal balance, the liquid rubidium would be stable kept in the cold end. Apparently, the lamp has no specialized heating circuit, all the heat of the bulb comes from the excitation circuit by RF power coupling and heat conduction. So we added a heat insulation structure to avoid excessive heat dissipation of the bulb.

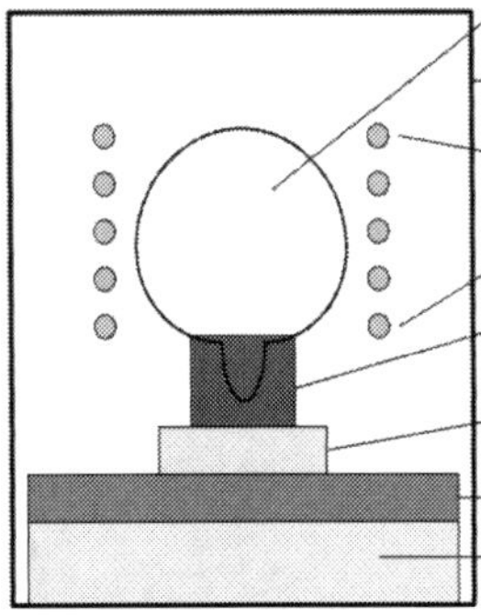

Figure 4. Structure diagram of the rubidium spectrum lamp originally designed.

Thermal simulation of the rubidium spectrum lamp was carried out by using ANSYS. The simulation result of the rubidium bulb is as shown in figure 5. The highest temperature of the bulb was about 117.53℃, and the lowest temperature was about 113.07℃. The rubidium bulb could form stable temperature gradient field, it means that the excess liquid rubidium atoms would finally flow to the cold end. However, the temperature gradient of the bulb was not steep enough, which limits the rate and stability of rubidium atoms redistribution. Besides, we can see that, most of the bulb body was at a high temperature. That might lead to the random moving of liquid atoms on the bulb body during the redistribution process, and result in frequency jumps of the RAFS, which is surely not desired.

Figure 5. Thermal simulation result of the rubidium bulb.

3.2 Optimization of Rb spectrum lamp

In order to make rubidium distribute more rapidly and smoothly, better temperature gradient field must be obtained, and the time for thermal balance should be also shortened. We redesigned the lamp as shown in figure 6.

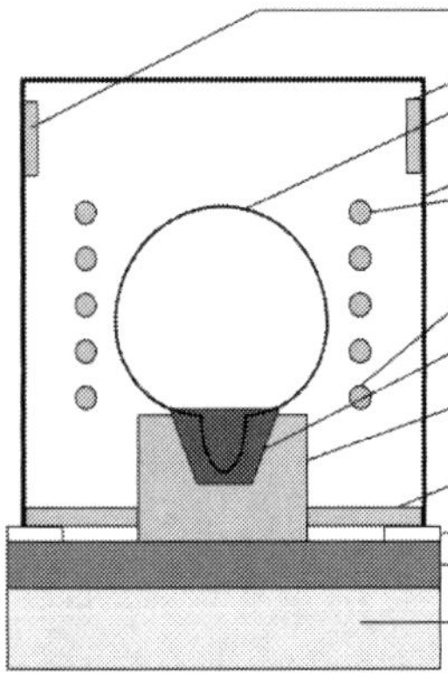

Figure 6. Optimized structure diagram of the rubidium spectrum lamp.

The optimization of rubidium spectrum lamp is mainly about the following aspects:

1)In the original design, all the heat of the bulb came from the excitation circuit, because we believe that single energy source has relatively lower noise and is better for more stable thermal balance and light intensity . However, that will take relatively more time for the whole lamp room to achieve stable thermal state and is not good for the lamp to obtain thermal balance rapidly. Aiming at this problem, we added a temperature control module(the heater and thermistor were installed in the specialized installation channels on the lamp room, while the control circuit was in a separate box for installing circuit system) to the lamp room to keep the aluminium lamp room at around 100℃. According to the simulation and test result of the original lamp, the bulb in the coil works at a temperature over 110℃, which is higher than the temperature of the lamp room we designed, so the added heating module would not influence the stable temperature of the bulb too much. The external constant temperature field applied to the lamp room would be conducive to rapid thermal balance of the spectrum lamp by transfer heat to the lamp through a heat conduction plate, and that would also be benificial to the quick-starting process of the lamp[10].

2)A better temperature gradient field needs a relatively colder end. So we removed the heat insulation structure shown in figure 4, and added an aluminium structure to help heat conduction of the cold end. The aluminium structure was connected to the lamp room through the heat conduction plate mentioned in the previous paragraph.

3)In the original design, the excitation circuit board was inside of the room to provide heat to the whole lamp. With the specialized heating module added, it is not necessary for the lamp to get too much heat from the circuit board. On the contrary, working under a high temperature for long time would have a negative effect on the excitation circuit[11]. So the lamp room was redesigned so that it was directly in contact with the heat conduction plate, and kept the excitation circuit board outside of the lamp room. Besides, a heat insulation ring

 https://doi.org/10.1051/matecconf/202031602003

with low thermal conductivity was added between the lamp room and the circuit board to avoid heat conduction.

Thermal simulation analysis of the optimized rubidium spectrum lamp module was developed using ANSYS, and the temperature gradient field of the rubidium bulb under stable operation is obtained in figure 7. During the stable operation of rubidium spectrum lamp, the highest temperature of rubidium bulb was about 120.26℃, and the lowest temperature was about 100.06℃. Compared with the original design, a better temperature gradient field was formed in the whole rubidium bulb (including the main body of the rubidium bulb) and the cold end had lower temperature. That is consistent with our design purpose, and beneficial for the liquid rubidium atoms to flow to the cold end rapidly and could also make the distribution of liquid rubidium more stable. According to the previous analysis of frequency jumps of RAFS, we believe that this optimization of rubidium spectrum lamp can effectively reduce the frequency jumps caused by the unstable rubidium distribution in the lamp.

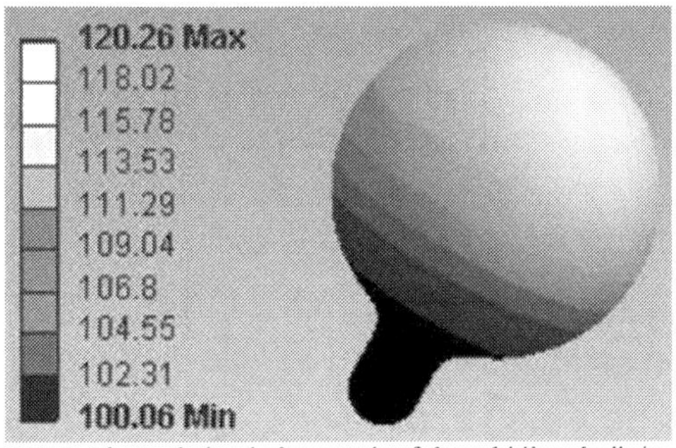

Figure 7. Thermal simulation result of the rubidium bulb in the optimized lamp module.

4 Experiments

According to our engineering experience and the operating data of on-board RAFSs, the process of rubidium steady-state distribution in a spectral lamp may be relatively long, such as several months and even more than 1 year, that means a frequency jump may occur in several days or even several months. So it is difficult to carry out frequency jumps observation experiments during the normal operation time of a RAFS. However, the start-up process of the RAFS is the main process for the liquid rubidium to achieve stable redistribution in the bulb, so the start-up process of the RAFS was selected for the experiment of different rubidium spectrum lamps test.

Two rubidium bulbs were selected from the same batch. The two rubidium bulbs have almost identical physical parameters and rubidium quantity, and their test performance was also similar under the same conditions. Then the rubidium bulbs were respectively installed in an original rubidium lamp module and an optimized rubidium lamp module, and the electrical parameters of the components were adjusted appropriately. The two lamps were successively tested in the same physics package. We started the frequency test at 10 minutes after the RAFS was locked, and the test results are shown in figure 8(In order to facilitate comparison of the two curves, the test data was normalized). The frequency test curve of the RAFS with the original designed rubidium lamp turned slowly during start-up, and there was also a slight frequency jump. This indicated that the process of liquid rubidium reaching stable distribution was slow and not smooth enough. On the other hand, the frequency of the RAFS with the optimized rubidium lamp module could quickly reach the stable state after starting up. The experimental result shows that the optimization of the thermal design for rubidium spectrum lamp can effectively promote the rapid and stable distribution of the liquid rubidium in the bulb. That will help the output frequency of the RAFS to reach the normal value quickly and eliminate the frequency jumps of the RAFS.

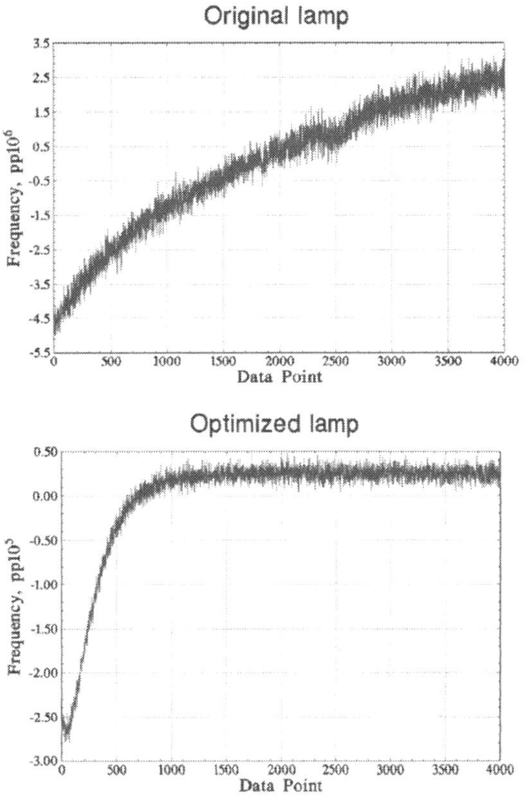

Figure 8. Frequency test of the two lamps during the start-up process.

5 Conclusions

The stability of the RAFS is directly related to the positioning accuracy of the navigation system and other satellites. By optimizing the design of rubidium spectral lamp, better temperature gradient was formed in the bulb to make the liquid rubidium in the bulb form a steady state distribution quickly and steadily, which would reduce the rubidium clock frequency jumps and help the output frequency of the RAFS to reach the normal value quickly. The study is benefit to improve the stability of the RAFS.

References

1. M. Epstein, *39th Annual Precise Time and Time Interval Meeting* (Beverly, MA, USA ,2007):59-78.
2. J. Camparo, *IEEE IELCONF*(2012):501-508.
3. B.Wang and J.P. Chen, *CPGPS 2017 Forum on Cooperative Positioning and Service*(Harbin, China,2017):258-261.
4. A. Khare, R. Arora, A. Banik, and S.D. Mehta, IEEE Transactions on Ultrasonics, Ferroelectrics, and Frequency Control, **63**, 2, (2016):326-335.
5. J.C. Camparo, *36th Annual Precise Time and Time Interval Meeting* (Washington,USA,2004):393-410.
6. J.Z. Cui, T. Yang, L. Zhang, Y.T. Liang, J.H. Tu, W. Yang, J.H. Zhang, Vac & Cry, **22**, 5(2016):271-274.
7. V. Formichella, *32nd URSI GASS* (Montreal, Canada, 2017).
8. S.G. He, F. Wang, F Zhao, H.H Wu, G.H. Mei, Acta Metrologica Sinica, **37**, 4(2016):433-436.
9. M. Huang, A Stapleton, and J. Comparo, *IEEE IFCS2018*(Olympic Valley,CA,USA,2018).
10. J.Q. Lian, J.H, Zhang, L.X Wei, *11st China Satellite Navigation Conference*(Chengdu,China,2020).(to be published)
11. C. Tao, J.L. Huang, L. Qin, J. Time & Freq,**38**, 1(2015) :8-12.

Simulation and Experimental Research on a Gas Liquid Separator with Rotary Impeller

Baihui Wang[1,2], Yingbin Li[2,a], Mingjun Deng[2], Baolu Shi[2], Wenjin Shang[2], and Bo Xu[2]

[1]*College of Aerospace Science and Engineering, National University of Defense Technology, Changsha 410005, China*
[2]*China Astronaut Research and Training Center, Beijing 100094, China*

Abstract. Gas-liquid separation technology under microgravity is the basis for various gas and liquid treatments on a manned spacecraft, which has a wide range of applications in Environmental Control and Life Support System. Dynamic gas-liquid separator is commonly used for the separation of gas-liquid two-phase flow, which has two essential performance parameters called liquid outlet pressure and separating efficiency. Predicting these two parameters accurately under a specific structure has guiding significance for design and application of the dynamic gas-liquid separator. In this study, CFD simulations were conducted using the Volume of Fluid (VOF) model at steady state conditions. In addition, experiments were designed to verify the accuracy of numerical results. Finally, the performance of the separator under microgravity was predicted. It is showed that the simulation method can be utilized to determine the transport performance of dynamic gas-liquid separator, which has significant value in design and optimization.

1 Introduction

Environmental Control and Life Support System (ELCSS) is the key technology to ensure that astronauts live and work in space environment. Gas-liquid separation is widely used in ECLSS, such as the recovery and treatment of sanitary wastewater, urine and condensate water. Gas-liquid separation technology under microgravity is divided into static gas-liquid separator and dynamic gas-liquid separator according to whether there are moving parts. Compared with static separation, the advantage of dynamic separation is that it can also provide partial pipeline circulation pressure after separation. There is centrifugal motion inside the dynamic separator which is generated by the rotation of gas-liquid mixture in a certain way. Separation will be achieved by centrifugal acceleration and density difference between gas and liquid. It is a key technology to separate gas-liquid mixture by rotary impeller.

The first generation of Mostly Liquid Separator was developed by USA in 1995, which is the application of dynamic separator with rotary impeller for the first time[1]. Samplatsky and Dean[2] developed Rotary Separator Accumulator for electrolytic oxygen system in 2002. Murdoch et al.[3] developed Rotary Drum Separator with built-in impeller for Sabatier CO_2 reduction system.

The designing difficulty of gas-liquid separator with rotary impeller (GLSRI) is how to determine structure parameters of separating chamber and impellers. In this study, CFD Simulation of GLSRI was carried out, and liquid outlet pressure and separating efficiency were investigated. In the next step, accuracy of the simulation was validated by experimental data with the model machine.

2 Numerical simulations

The separation and recovery of condensate is one of the applications of GLSRI. The schematic of condensate collection is shown in Fig.1, where No.1 is a suction pump, which is used to transport the mixture of condensate and air upstream to No.2; No.2 is the GLSRI, which is used to separate condensate and air, and air will be discharged directly into the atmosphere; No.3 is a pressure sensor, which is used to detect the pressure at the liquid outlet of No.2; No.4 is a solenoid valve, which is used to control the condensate pressure and upstream flow, and No.5 is a storage tank, which is used to collect the separated condensate.

Technical indexes of GLSRI are presented in Table 1, in which output pressure of the liquid outlet must be no less than 20 kPa. Meanwhile, the volume fraction of gas in separated condensate must be at most 1 %.

Figure 1. Schematic of condensate water collection.

[a] Corresponding author: liyingbin507@139.com

Table 1. Technical indexes.

Parameter	Index
Components	Mixture of air and condensate
Flow	1~2 L/min, about 50% is air
Air volume	Less than 1 %
Outlet Pressure	No less than 20kPa
Motor speed	(1150±50) rpm

2.1 Geometry modelling and mesh generation

The fluid domain model was generated from the preliminary design of the structure with a separation chamber and rotary blades according to the technical indexes in Table 1, which is shown in Fig.2(a). Four vertical rotary impellers were adopted, of which the diameter is 140 mm, and the width is 30 mm. Moreover, the chamber can collect 500 ml of water. As shown in Fig.2(b), mesh generation of the fluid domain model was achieved using unstructured grid elements, and the number is 597724.

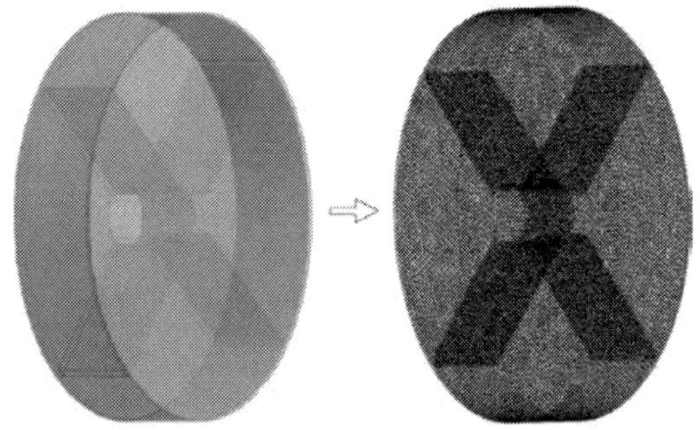

(a) model (b) mesh

Figure 2. Fluid domain model and mesh.

2.2 Numerical method

2.2.1 Physical models specification

VOF model, Mixture model and Euler model are commonly used for multiphase flow analysis. Meanwhile, discrete phase model works when it comes to the calculation of particle flow problems. For the current study, VOF model is selected which can obtain the interface between multiple non-soluble fluids[4].This characteristic can be well used to determine the separation efficiency of GLSRI. In the VOF model, different fluid components share one continuous equation and one momentum, shown in Eq.(1) and Eq.(2).

$$\frac{\partial \rho}{\partial t}+\nabla\cdot\left(\rho\vec{u}\right)=Smu \quad (1)$$

$$\frac{\partial\left(\rho\vec{u}\right)}{\partial t}+\nabla\left(\rho\vec{u}\vec{u}\right)=-\nabla p+\nabla\left(\mu\left(\nabla\vec{u}+\nabla\vec{u}^{T}\right)\right)+\rho\vec{g}+\vec{T}_{\sigma} \quad (2)$$

where ρ is the mixture density; μ is the mixture viscosity; $\vec{u}$ is the mixture velocity; $\vec{T}_{\sigma}$ is the surface tension. The volume fraction occupied by each element of the whole fluid domain.

The surface tension between condensate and air was considered by using the tension model of continuum surface force, and the surface tension coefficient was regarded as 0.073 N/m. The continuous equation was solved by implicit scheme, which is not limited by courant number and can be solved in steady-state conditions. To prevent diffusion problems caused by low mesh quality, interface Anti-Diffusion should be taken into account. Due to the large body force generated by the rotary impeller, implicit body force is a key factor to improve astringency of the model.

In this study, SST k-ω turbulence model was used in the simulations, which combines k-ε model and k-ω model by weighted average method, using k-ω model at the wall surface and k-ε model selecting k-ω model away from the wall surface. In this way, the dependence of ω on the relative value of free flow outside the boundary layer in k-ω model is overcome[5].

2.2.2 Boundary conditions and solution methods

The rotary impeller divides the fluid domain into two parts, that is, the fluid domain that the impeller drives to rotate and the stationary fluid domain at the gap between the rotating fluid domain and the wall. Sliding grid method can be used for the simulation of rotary fluid domain by determining the rotation axis and the speed. The boundary of the rotary fluid domain was modelled as moving wall, of which the relative speed was assumed to be 0. The remaining wall boundaries were modelled as non-slip walls, and standard wall function method was used to determine the flow near the wall.

In order to determine liquid outlet pressure and separating efficiency, the simulation was carried out in steady-state mode under different working conditions. Coupled algorithm was used for pressure-velocity coupling. Momentum, turbulent kinetic energy and specific dissipation rate were all discrete by using second-order upwind scheme while volume fraction was discrete by using the compressive format.

3 Simulation results

When GLSRI is working, the thickness of the liquid increases along the circumference with the continuous injection of condensate until the chamber is filled with condensate. Meanwhile, gas flows out through the gas outlet into the atmosphere. In this study, condensate was replaced by pure water. First, the case that 50 ml of water injected into the chamber was calculated. Then the case of adding another 50 ml of water was recalculated until the chamber was filled with water.

3.1. Liquid outlet pressure

The pressure distribution cloud diagrams with different water volume inside the chamber is shown in Fig.3, where the impeller rotates anticlockwise at 1150 rpm. It is observed that the pressure near the rotation axis is low, and the pressure near the circumferential boundary is much higher, especially the edge of impeller. The pressure slightly differs between two symmetry points due to the influence of gravity, and the pressure near the gravity is much higher.

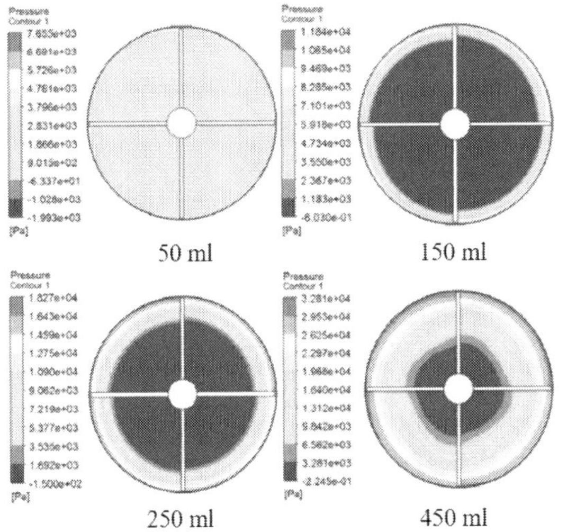

Figure 3. Liquid pressure distribution.

The pressure at the liquid outlet under different water volume is presented in Table 2. It is indicated that the liquid outlet pressure is over 20 kPa, which meets the requirement in Table 1 when the water injected reaches 350 ml. Therefore, every time GLSRI works, the volume of condensate in the mixture cannot be less than 350 ml.

Table 2. Simulation results of liquid outlet pressure.

Water volume (ml)	Pressure (kPa)
50	1.339
100	2.880
150	6.635
200	10.099
250	13.583
300	17.224
350	20.919
400	24.476
450	28.144
500	32.128

3.2 Separating efficiency

The separating efficiency of GLSRI is defined as the volume fraction of air in liquid at the liquid outlet. The air volume fraction cloud diagrams with different water injected into the chamber are shown in Fig.4. It is observed that water and air layering is obvious under the effect of the centrifugal force generated by the rotary impeller, indicating that separation of gas-liquid two-phase flow can be effectively achieved by this impeller structure. However, under the agitation of the rotary impeller, there is still some air distributed in water. As a result, the gas-liquid mixture cannot be separated completely.

When the water volume in the chamber is 50 ml, the liquid outlet is not covered by water completely, so the separating efficiency at this volume was not considered. According to Fig.4, the separating efficiency with different water volume is plotted in Fig.5. It is illustrated that the separating efficiency at each point is lower than 0.6 %, which satisfies the requirement of not more than 1 % (shown as the red line in Fig.5) mentioned in Table 1.

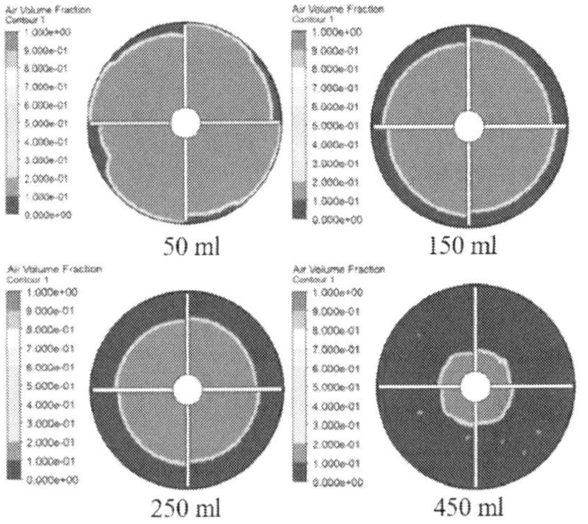

Figure 4. Air volume fraction distribution.

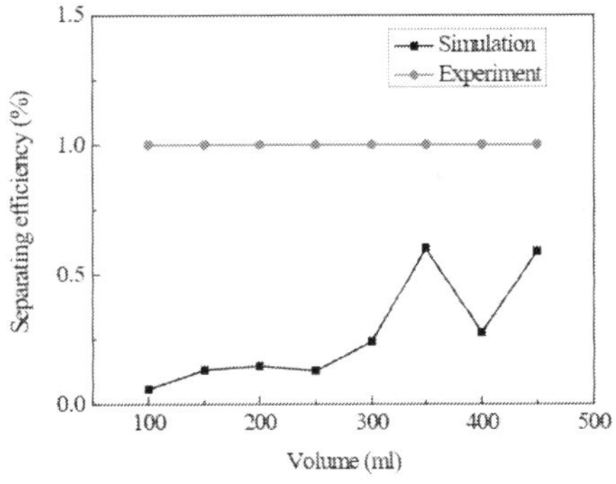

Figure 5. Separating efficiency by simulation.

4 Experimental verification

According to the simulation results in Section 3, the preliminary design of GLSRI[6] can meet the technical indexes in Table 1. Therefore, an experimental prototype was designed to verify the accuracy of simulation results. The performance test platform of GLSRI is shown in Fig.6, and the prototype used in experiments is shown in Fig.7.

Figure 6. Performance test platform.

Figure 7. Experimental prototype.

4.1 Test experiment of liquid outlet pressure

50 ml of pure water was injected into separating chamber. Record the value of pressure sensor after the system runs steadily. Increase by 50 ml per time until the volume of water in the chamber reaches 500 ml, during which the gas outlet is always open. The liquid outlet pressure values under these 10 water storage states are listed in Table 3.

Table 3. Experimental data of liquid outlet pressure.

Water volume (ml)	Pressure (kPa)
50	0.59
100	3.04
150	6.88
200	10.08
250	13.76
300	17.28
350	20.96
400	24.48
450	28.32
500	31.52

Experimental data in Table 3 are compared with simulation results in Table 2 as shown in Fig.8. It is observed that the pressure of liquid outlet increases gradually with the increase of water volume, which can be regarded as a linear increase. When the water volume is 50 ml, the deviation between simulation results and experimental data is large. This is because water volume is not enough to cover the liquid outlet completely, which results in a large fluctuation in the reading of pressure sensor. The error of other points is plotted in Fig.9, in which it can be seen that the maximum error is still less than 6 % and the minimum error is only 0.014 %. Moreover, the rising tendency of liquid outlet pressure is the same between simulation results and experimental data. Therefore, it can be illustrated that the CFD simulation of GLSRI in this paper is real and effective.

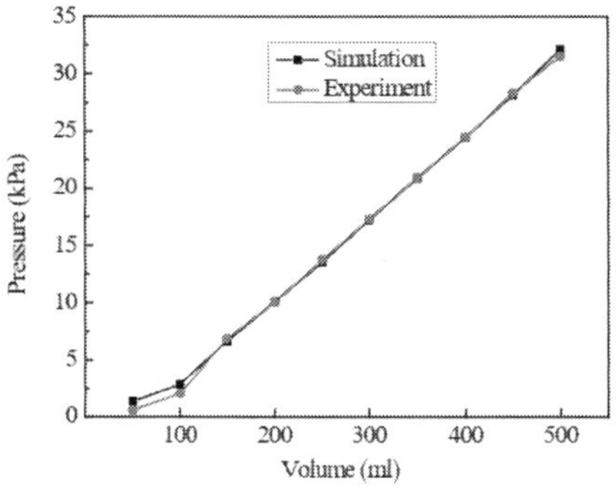

Figure 8. Comparison of liquid outlet pressure.

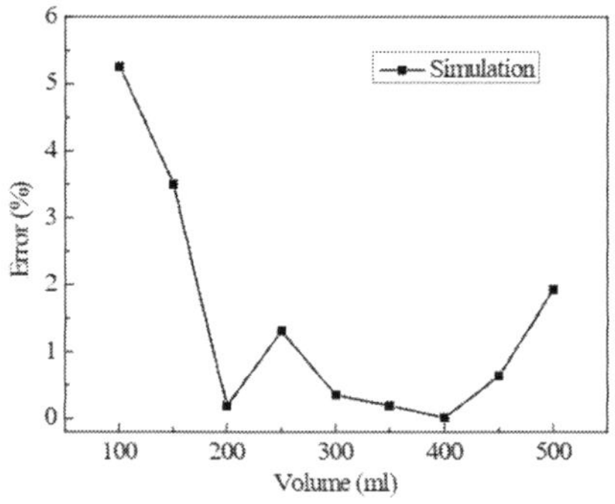

Figure 9. The error of each point.

According to the simulation, the performance of GLSRI under microgravity can be determined, and the relationship between the liquid outlet pressure and water volume was carried out as shown in Fig.10, in which the pressure of liquid outlet was almost the same as that under gravity.

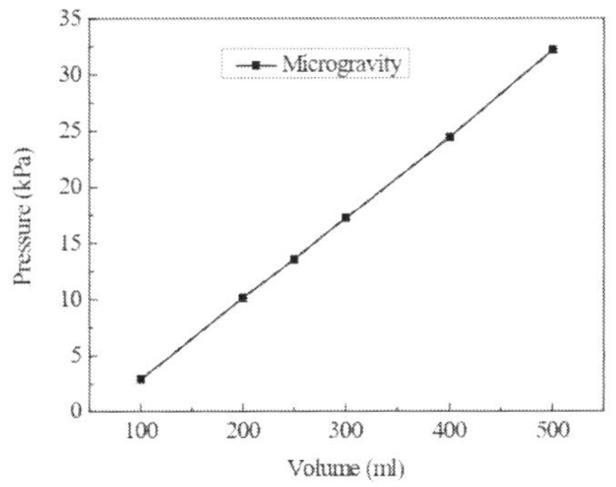

Figure 10. Liquid outlet pressure under microgravity.

4.2 Test experiment of separating efficiency

The important components of experimental apparatus are a prototype, a pressure sensor and a measuring cylinder. The schematic of separating efficiency test experiment is shown in Fig.11, where No.1 is a cup for supplying water, whose flow is 2 L/min; No.2 is a gas-liquid mixer for producing gas-liquid mixture; No.3 is a prototype for the separation of water and air; No.4 is a pressure sensor used for detecting the liquid outlet pressure; No.5 is an electromagnetic valve, the switch of which is controlled by No.5; No.6 is a measuring cylinder, which is used to measure the volume of bubbles in water flowing from the liquid outlet; and No.7 is a power supply.

In order to meet the requirement of separating efficiency, it is necessary to ensure that the liquid outlet is covered by water when opened and closed. According to the liquid pressure curve in Fig.8, during the experiment, when liquid outlet pressure reaches 15 kPa, No.5 is open for separation, and when the pressure drops to 7.5 kPa, No.5 is closed. Repeat the experiment for several times. Read the bubble volume collected in No.6 to get the separating efficiency. The final average separating efficiency is 0.04 %, which meets the requirement that the separating efficiency must be less than 1 %, and the accuracy of the CFD simulation is further proved by comparison with Fig.5.

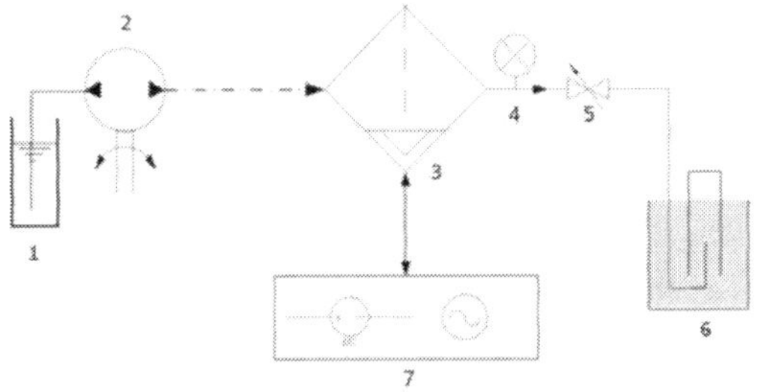

Figure 11. Schematic of separating efficiency test.

5 Conclusions

In this paper, a GLSRI that can be used for the separation and recovery of condensate in a manned spacecraft was designed. The internal gas-liquid two-phase flow separation was simulated under steady state conditions combined with the VOF multiphase flow model and sliding grid technology.

The relationship between the liquid outlet pressure and the water volume inside the chamber was obtained. The pressure of liquid outlet increases linearly with the increase of water volume. The liquid outlet pressure exceeds 20 kPa when the water volume reaches 350 ml.

The relationship between the separating efficiency and the water volume inside the chamber was obtained. As long as the liquid outlet is covered by water completely, the separating efficiency is less than 1 %.

The simulation method is proved to be effective comparing the simulated liquid outlet pressure with experimental results whose maximum error is less than 6 %. The comparison of separating efficiency can further illustrate that the simulation method can be used for the prediction of the performance when GLSRI works. And then the performance of GLSRI under microgravity was determined.

As a result, the information get on the ground can be used to evaluate the performance of GLSRI in space, which is significant for the design and application of GLSRI in different situations on a manned spacecraft. In the future work, the separating of gas-liquid mixture under transient conditions will be researched on the basis of this simulation and also the influence of different factors on the performance of GLSRI.

Acknowledgements

This work was supported by China Astronaut Research and Training Center under project SYFD061909.

References

1. A.W. Lanzarone, W.C. Dean, S.G. Howard. SAE Technical Paper. No.972374 (1997)
2. D.J. Samplatsky, and W.C. Dean. SAE Technical Paper. No. 2002-01-2360 (2002)
3. K. Murdoch, J. Fort, M. Barone, and D. Holder. SAE Technical Paper. No. 2005-01-2863 (2005)
4. P. Hugo, et al. Journal of Petroleum Science and Engineering, **139**, 49-61 (2016)
5. S. Derakhshan, F. Riahi, M. Bashiri. SPE Production & Operations, **33(02)**, 320-335 (2018)
6. F. Kulor, E.D. Markus, and M.W. Apprey. International Journal of Mechanical Engineering and Robotics Research, **8**, 639-645 (2019)

Modeling and analysis of regenerative cooling channels for scramjet engine

Feng Gao, Qian Zhang[a], Hongyu Xiao, Chengtao Zhang, Xuefeng Xia

Air and Missile Defense College, Air Force Engineering University, 710051 Shaanxi Xian, China

Abstract. This paper presents a CFD simulation analysis method for the regenerative cooling channel of a supercombustion ramjet engine, and performs three-dimensional modeling and analysis. The influences of inlet mass flow, inlet pressure, and inlet temperature on the flow and heat transfer characteristics of kerosene in the cooling channel were analyzed, and the conclusions are as follows: The larger the inlet mass flow, the lower the maximum wall temperature and oil temperature. The change trend of wall temperature is basically the same under different inlet pressure conditions, all of which increase violently first, then stabilize and then suddenly increase suddenly, then decrease almost to the outlet, and the temperature is almost the same under different conditions.As the inlet temperature decreases, the temperature difference between the wall surface and kerosene becomes larger, and the convective heat transfer coefficient gradually increases.

1 Introduction

With the increasing number of flying Mach numbers, the thermal protection problem of hypersonic vehicles becomes more and more serious. The work of super-combustion stamping engines is in extreme harsh environments such as high speed, high heat flux density, high-intensity combustion, etc., and a feasible and reasonable thermal protection is urgently needed. Due to the particularity of the working environment of the combustion chamber, the commonly used combustion chamber cooling methods include film cooling, ablation cooling, perspiration cooling, and regenerative cooling [1]. Compared with other types of cooling methods, regenerative cooling has great advantages. At present, the more successful hypersonic aircrafts in the world such as X-43A, X-51A and other super-combustion stamping engines use regenerative cooling technology. A key technology in supersonic vehicle technology. In recent years, with the continuous maturity of computer technology, numerical simulation has also been widely used, and the research on regenerative cooling technology at home and abroad has also achieved considerable achievements [2-10]. However, most scholars have simplified the cooling channel based on the criterion relationship or based on some flow hypothesis to build a simple one-dimensional heat transfer model. There are relatively few numerical simulations of three-dimensional channels, and most of them have studied the heat transfer characteristics of the coolant in the channels, and there is less research on the effect of the inlet conditions of the coolant on the cooling efficiency. The channel was modeled by 3D CFD, and calculated and analyzed. The flow and heat transfer characteristics of kerosene in the channel under different inlet mass flow, inlet pressure, and inlet temperature are analyzed.

2 Calculation models and numerical methods

2.1 Geometri cmodel and basic assumptions

A super-combustion stamping engine using a regenerative cooling channel is welded from different heat exchange panels. The simple structure is shown in Figure 1. The schematic diagram of the heat exchange plate with regenerative cooling channel is shown in Figure 2.

[a] Corresponding author: zhangqian8913@163.com

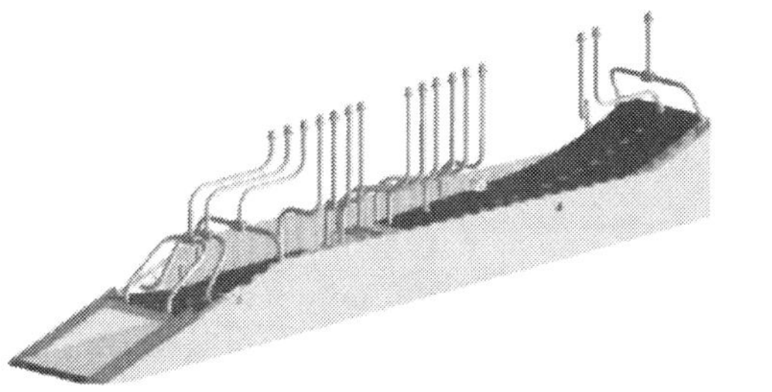

Figure 1. Scramjet engine with regenerative cooling.

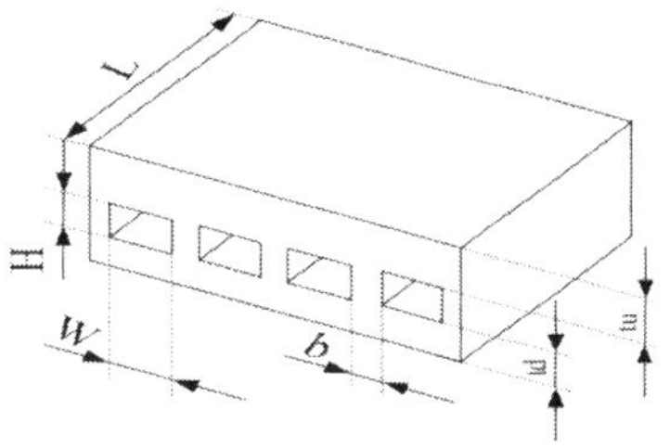

Figure 2. Heat exchange plate for regenerative cooling channel.

Due to the symmetry of the regenerative cooling channel, in order to simplify the calculation, only one half of the flow unit of the cooling channel is analyzed. Figure 3 is a cross-sectional view of the half of the flow unit of the regenerative active cooling channel. Where a is half the rib width, b is the height of the cooling channel, c is the half width of the channel, d is the thickness of the outer wall, and e is the thickness of the inner wall. The channel material is made of alloy steel, and the coolant is made of domestic aviation kerosene. Its physical properties are given in [11].

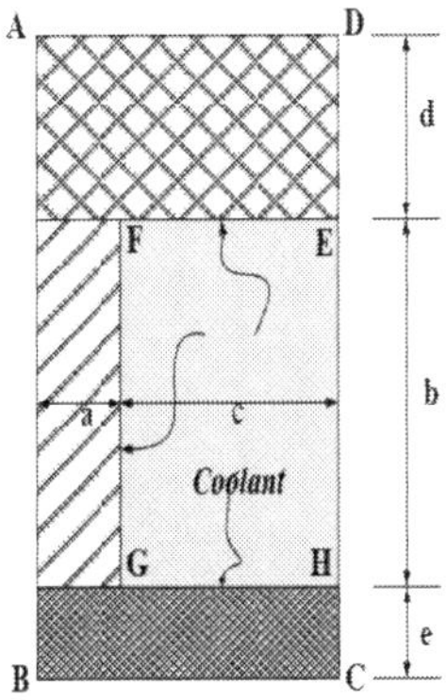

Figure 3. Sectional view of flow unit of regenerative active cooling channel flow unit.

Due to the complexity of heat transfer in regenerative active cooling channels, the following assumptions are made on the theoretical model:

(1) The fluid in the channel has no phase change latent heat release and absorption. The heat transfer process has the characteristics of unidirectional flow, and the fluid is in a supercritical state in the channel;

(2) The cross-sectional shape of the channel is regarded as a rectangle;

(3) The thermal resistance between the connections is negligible;

(4) The symmetry plane of the side ribs is an adiabatic surface;

(5) Do not consider the influence of radiative heat transfer;

(6) Dissipate heat from the channel to the environment;

(7) Do not consider the carbon deposits of aviation kerosene.

2.2 Numerical calculation method

2.2.1 Governing Equation

The governing equations include mass continuity equations, energy equations, and momentum equations, as shown in equations (1), (2) and (3), respectively.

$$\frac{\partial \rho}{\partial t} + \frac{\partial}{\partial x_i}(\rho u_i) = 0 \quad (1)$$

$$\frac{\partial}{\partial t}(\rho E) + \frac{\partial}{\partial x_i}(u_i(\rho E + p)) = \frac{\partial}{\partial x_i}(k_{eff}\frac{\partial T}{\partial x_i} - \sum_{j'} h_{j'} J_{j'} + u_j(\tau_{ij})_{eff}) + S_h \quad (2)$$

$$\frac{\partial}{\partial t}(\rho u_i) + \frac{\partial}{\partial x_j}(\rho u_i u_j) = -\frac{\partial p}{\partial x_i} + \frac{\partial \tau_{ij}}{\partial x_j} + \rho g_i + F_i \quad (3)$$

In the formula (2), k_{eff} is the effective thermal conductivity: $k_{eff} = k_l + k_t$, k_t is the turbulent thermal conductivity, which is determined based on the turbulence model. J_j is the diffusion flux of component j', S_h contains the heat of chemical reaction and other user-defined volumetric heat source terms, $E = h - \frac{p}{\rho} + \frac{u_i^2}{2}$.

In equation (3), ρg_i and F_i are the gravitational volume force and the external volume force in the i direction, respectively, and τ_{ij} is the stress tensor: $\tau_{ij} = [\mu(\frac{\partial u_i}{\partial x_j} + \frac{\partial u_j}{\partial x_i})] - \frac{2}{3}\mu\frac{\partial u_l}{\partial x_l}\delta_{ij}$. In the formula, μ is the viscosity coefficient: $\mu = \mu_l + \mu_t$, μ_l is the laminar viscosity coefficient, μ_t is the turbulent viscosity coefficient, which is defined according to the turbulence model.

2.2.2 Meshing and boundary conditions

Simulation conditions: The mass flow rate of the coolant aviation kerosene inlet is 0.01kg / s, the calculated area is length 0.5m, width 1.5mm, heat flow density of the inner wall $1\text{MW}/\text{m}^2$, and inlet pressure 4MPa. Channel parameters: The thickness of the half rib is 1mm, the channel is 4mm high, 4mm wide, the outer wall thickness is 3mm, and the inner wall thickness is 2mm.

In order to simplify the problem, the horizontal heat transfer between adjacent channels is not considered, and the symmetry is used to reduce the number of meshes and shorten the calculation time. The fluid-solid coupling interface is encrypted when the mesh is divided, compared with the number of 200,000 meshes and 600,000 meshes. As a result of the calculation, the temperature difference is within the range of 15 K, so the calculation uses 200,000 grids. The grid division is shown in Figure 4.

In numerical calculations, proper and correct boundary conditions are of great significance to the accuracy of the calculation. Different boundary conditions are chosen to obtain different results.

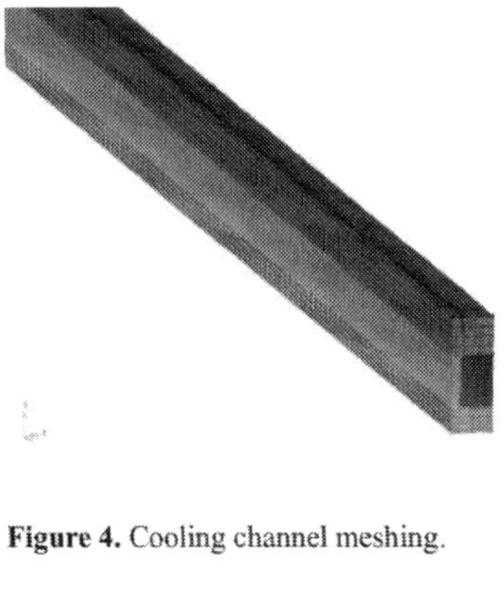

Figure 4. Cooling channel meshing.

The inlet AO is the boundary condition for the mass flow into the inlet, mass flow $\dot{m} = \dot{m}_0$, inlet pressure $p = p_0$, inlet temperature $T = T_0$, turbulent kinetic energy $k = k_0$, and turbulent dissipation rate $\varepsilon = \varepsilon_0$, where k_0 and ε_0 are shown by equations (4) and (5) respectively.

$$k_0 = \frac{3}{2}[\bar{u}(0.16Re_{DH})^{-1/8}]^2 \quad (4)$$

$$\varepsilon_0 = C_u^{3/4}\frac{k^{3/2}}{l} \quad (5)$$

Where $\bar{u}$ is the average fluid velocity, DH is the hydraulic diameter, C_u is the empirical constant, 0.09, and l is the turbulence length scale, which is approximately 0.07D.

The upper wall surface, the outlet wall and the inlet wall surface are adiabatic wall surfaces, and the wall surface heat flow density is zero. The lower wall surface is the wall surface of the heating section, and the heat flux density is the heat flux density applied in the physical model. Both sides are set to symmetrical boundary conditions, and the axial velocity, temperature, and pressure are all zero. The outlet cross section is a pressure outlet, and the outlet back pressure is set to zero.

2.2.3 Solving the governing equations

The governing equation is solved according to incompressible and variable physical conditions. Fluent was used for numerical simulation. The RNG $k-\varepsilon$ model was used for the turbulence model. The equation was discretized using the second-order upwind style. The SIMPLEC algorithm was used to solve the coupling of pressure and velocity. All variables in the governing equation have second-order accuracy in time and space.

3 Calculation results and analysis

3.1 Impact of inlet mass flow

In order to analyze the effect of different mass flow on the heat transfer performance of the cooling channel, this section only changes the kerosene inlet mass flow without changing the channel geometry, inlet temperature and inlet pressure, and simulates the situation of 10 different mass flow inlets. The parameters and results are as follows:

Table 1. Comparison of calculation results of different inlet mass flows

Study No.	Quality flow (g/s)	Inlet temperature (K)	Inlet pressure (MPa)	Maximum wall temperature (K)	Maximum oil temperature (K)
1	4	300	4	1254.927	1010.276
2	6			1062.356	900.240
3	8			905.512	813.661
4	10			825.071	748.982
5	15			810.710	699.982
6	20			801.235	680.021
7	25			792.001	672.104
8	30			770.842	653.943
9	40			762.570	640.009
10	50			738.993	633.182

It can be seen from Table 1 that as the inlet mass flow rate increases, the maximum wall temperature and maximum oil temperature are decreasing. In order to more clearly analyze the relationship between the inlet mass flow rate and the maximum wall temperature and maximum oil temperature, the inlet mass flow rate is taken as the abscissa and the ordinate is the temperature, as shown in FIG. 5.

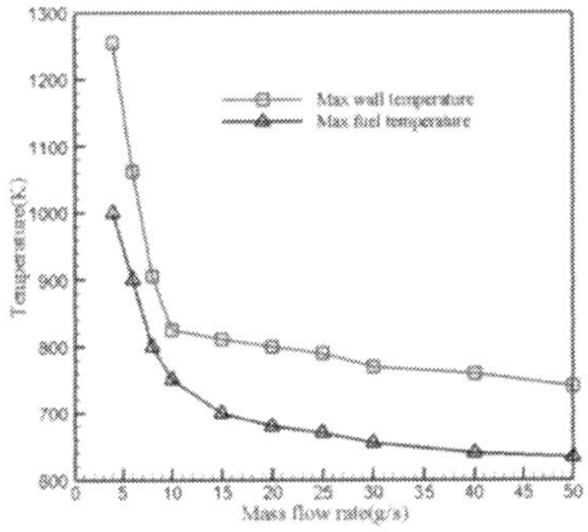

Figure 5. Temperature comparison of different inlet mass flows

It can be found from Figure 5 that the larger the inlet mass flow, the lower the maximum wall temperature and oil temperature. When the inlet mass flow rate is less than 8g / s, the maximum wall temperature and oil temperature rise sharply as the mass flow rate decreases.When the inlet mass flow rate is greater than 20g / s, the maximum wall temperature and oil temperature decrease as the mass flow rate increases. Slower. Considering that the higher the mass flow rate is, the larger the flow rate is under the condition of constant cross-sectional area. When the flow velocity is too large, the viscosity influence is large, the flow resistance is also large, and the kinetic energy loss of kerosene in the channel is faster. Therefore, the inlet mass flow rate is more suitable between 8g / s -20g / s.

3.2 Influence of inlet pressure

Figures 6 and 7 show the distribution of wall surface temperature and central flow temperature at the centerline of the lower wall surface of the cooling channel with the channel under different inlet pressure conditions. The change trend of wall temperature is basically the same under the three inlet conditions. All of them increase sharply, then stabilize and then abruptly increase, and then decrease near the outlet. The temperature is almost unchanged, and the temperature of the central flow is not much different under different conditions. Consistently showing a linear growth trend, the temperature difference is not greater than 5K under different inlet conditions.

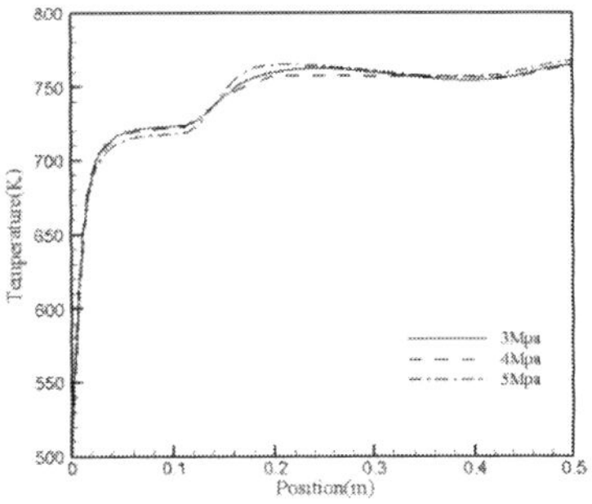

Figure 6. Distribution of temperature at the centerline of wall with channel.

https://doi.org/10.1051/matecconf/202031603002

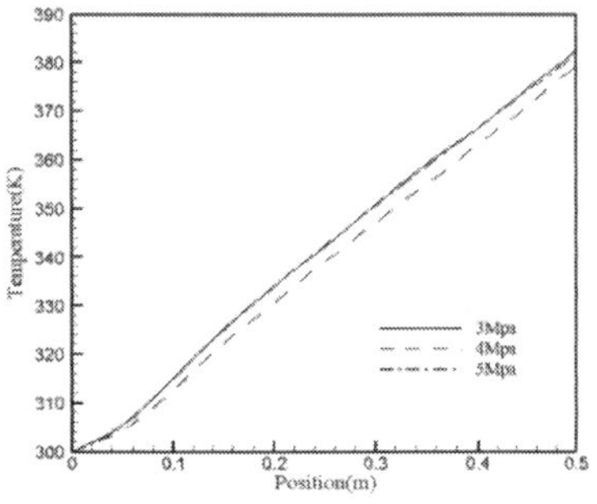

Figure 7. Distribution of central flow temperature with channel.

When the inlet flow rate is 10 g / s and the inlet temperature is 300 K, the cases where the inlet pressures are 3 MPa, 4 MPa, and 5 MPa are analyzed. Three cross sections in the channel (inlet cross section z = 0m, center plane z = 0.25m, outlet section z = 0.5m) was analyzed for temperature distribution, as shown in Figure 8、Figure 9 and Figure 10.

Comprehensive analysis shows that under different inlet pressures, the temperature difference is small. This is because under the condition of the same inlet mass flow rate and constant inlet temperature, an increase in pressure will cause an increase in kerosene density and dynamic viscosity, resulting in a decrease in flow velocity and Reynolds number. Therefore, pressure has little effect on the temperature of kerosene in the channel.

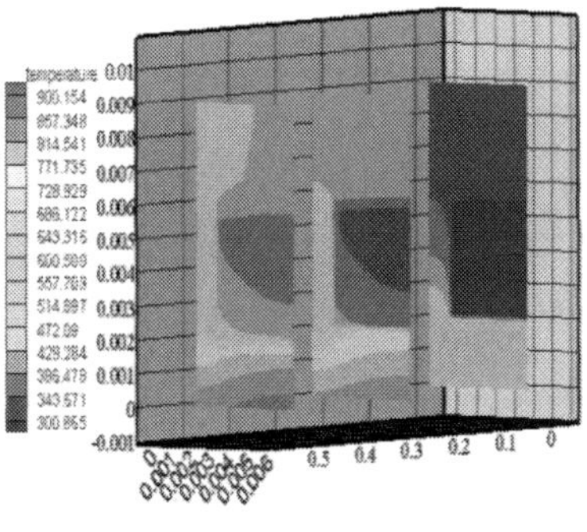

Figure 8. Cloud map of channel temperature distribution at 3MPa.

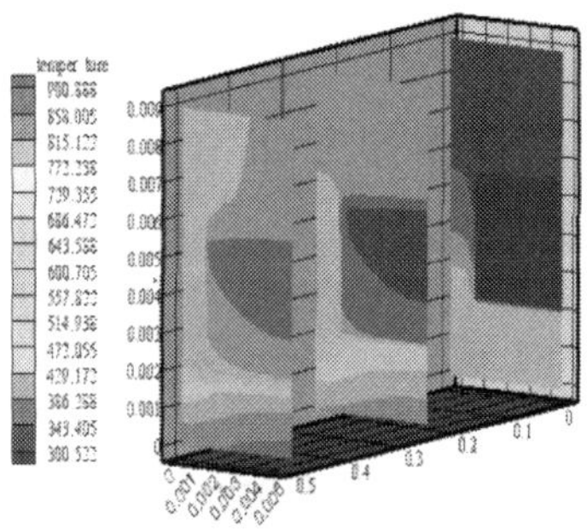

Figure 9. Cloud map of channel temperature distribution at 4MPa.

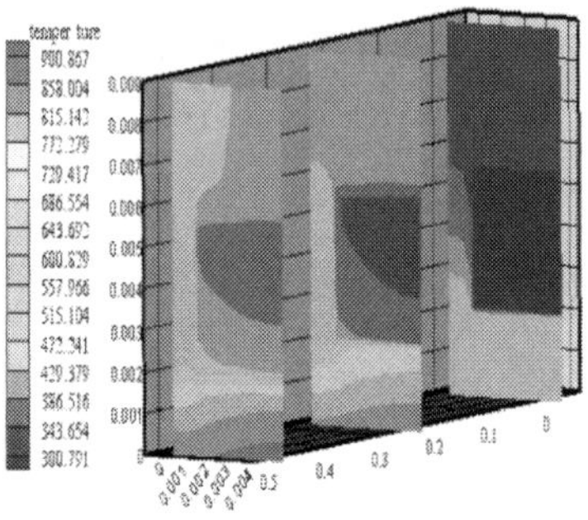

Figure10. Cloud map of channel temperature distribution at 5MPa.

3.3 Influence of inlet temperature

In order to study the influence of the inlet temperature on the heat transfer performance of the cooling channel, this section analyzes the cooling channel when the inlet temperature is 200K, 300K, 400K, and 500K when the inlet mass flow is 0.01kg / s and the inlet pressure is 4MPa. Convection heat transfer situation.

Figures 11 to 14 show the distribution of the centerline wall temperature, center flow oil temperature, kerosene pressure, and heat transfer coefficient along the channel at different inlet temperatures. It can be seen from Fig. 11 that when the inlet temperature is 300K, 400K, and 500K, the wall temperature of the inner wall of the inner wall varies, and the wall temperature suddenly rises at 0.11m, 0.19m, and 0.31m, respectively, and the heat transfer deteriorates. Then it decreases, and the change near the exit section is small. When the inlet temperature is 200K, the wall temperature does not change abruptly. This is because the temperature of kerosene is low and cannot reach the temperature at which kerosene deteriorates, and no heat transfer deterioration occurs. It can be seen from FIG. 12 that the temperature change trend of the central flow is basically the same, and both gradually increase along the axial direction. The influence of inlet temperature on the pressure of kerosene is mainly reflected at the entrance. As can be seen from Figure 13, when kerosene flows in the channel, the pressure gradually decreases along the channel. The kerosene pressure is almost the same at the temperature of 200K and 300K, but only at the entrance. The kerosene pressure is slightly higher at a temperature of 300K. When the inlet temperature is high, the corresponding pressure is large, and the pressure loss of kerosene in the channel is also large. When the inlet temperature is low, the pressure loss is small. Figure 14 shows the change of kerosene heat transfer coefficient along the channel. As a whole, the inlet temperature has a greater impact on the heat transfer coefficient. As the inlet temperature decreases, the temperature difference between the wall and kerosene becomes larger, and convective heat transfer The coefficient gradually increases. Observe the curves when the inlet temperature is 300K, 400K, and 500K. When the wall temperature rises sharply during the inlet stage, the heat transfer coefficient decreases sharply. When the wall temperature does not change much, the heat transfer coefficient also remains almost unchanged. When the wall temperature changes suddenly, The heat transfer coefficient decreases again, and finally the wall temperature tends to be constant, and the change of the heat transfer coefficient is relatively small. There is no deterioration of heat transfer on the wall surface at the inlet temperature of 200K, and the heat transfer coefficient only shows a sharp decline and then a gentle decline.

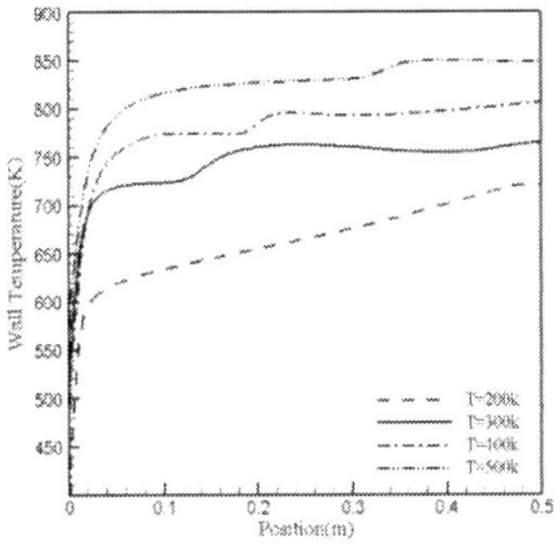

Figure 11. Distribution of wall temperature along the centerline of the inner wall.

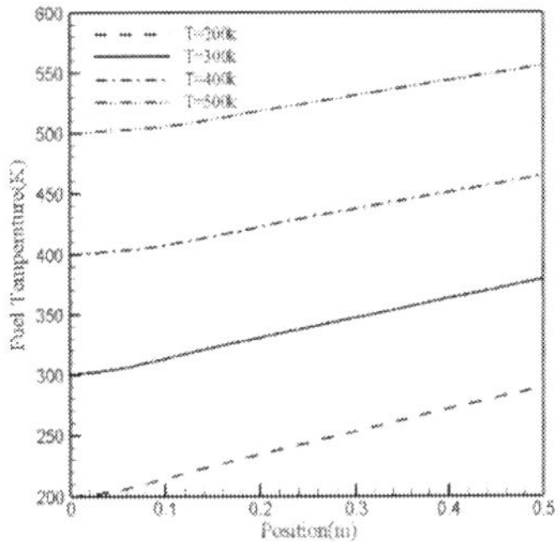

Figure 12. Distribution of central flow oil temperature along the channel.

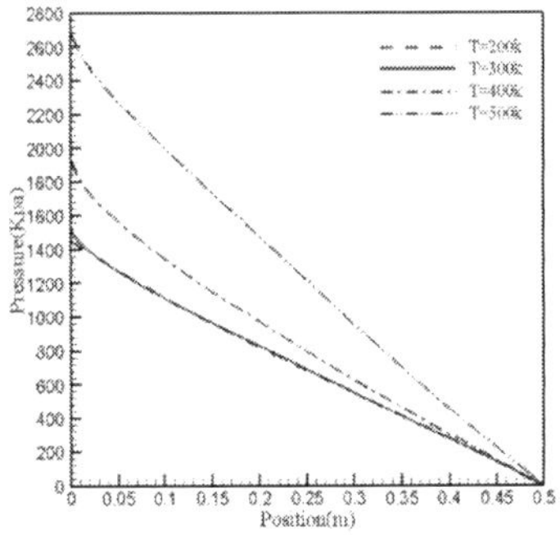

Figure 13. Distribution of central flow kerosene pressure along the channel.

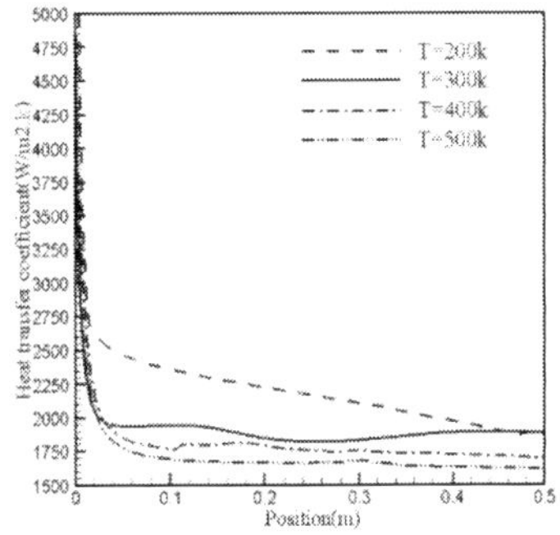

Figure 14. Distribution of heat transfer coefficients along the channel.

For a more intuitive analysis of the cooling channel and the distribution of kerosene temperature in the channel, Figures 15 to 18 show three cross-sections in the channel (inlet cross-section z = 0m, center plane z = 0.25m, outlet cross-section z=0.5m) at four inlet temperatures. Because the lower wall surface of the cooling channel contacts the engine combustion chamber, the temperature is much higher than the upper wall surface, and the temperature difference is about 400K. There is a large heat flow near the engine combustion chamber side, which causes a large temperature change. The temperature distribution has a more pronounced rib effect. Similarly, kerosene has a large temperature gradient near the lower wall surface and the fin side. With the increase of the inlet temperature, the temperature of the kerosene in the channel increases, the temperature of the lower wall gradually increases, and the temperature of the lower wall changes greatly at different inlet temperatures, while the temperature of the upper wall changes less. The smaller the inlet temperature, the larger the temperature difference between kerosene and the lower wall surface, and the better the heat transfer performance.

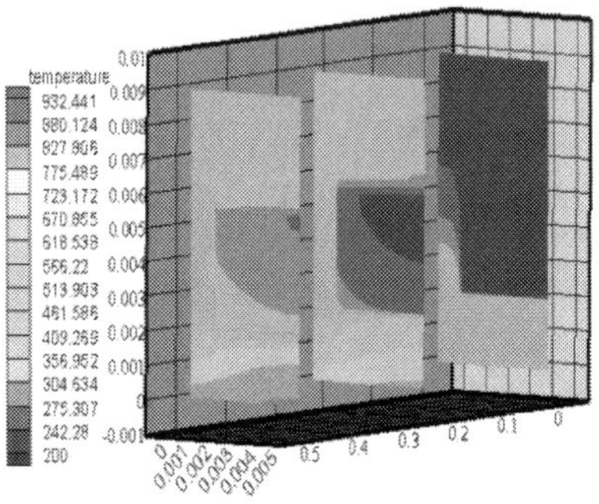

Figure 15. Cloud diagram of channel temperature distribution at inlet temperature of 200K.

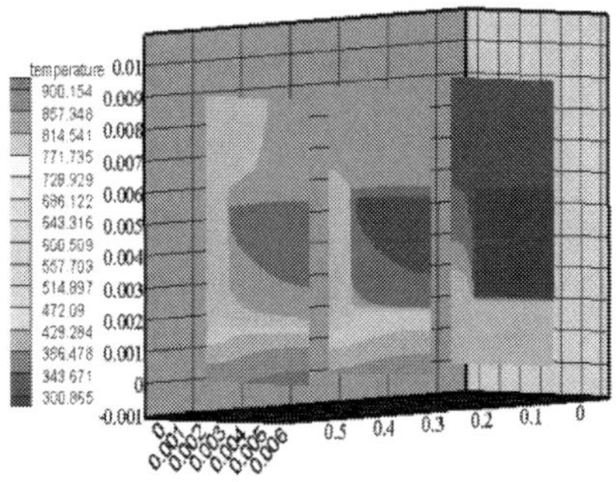

Figure 16. Cloud diagram of channel temperature distribution at inlet temperature of 300K.

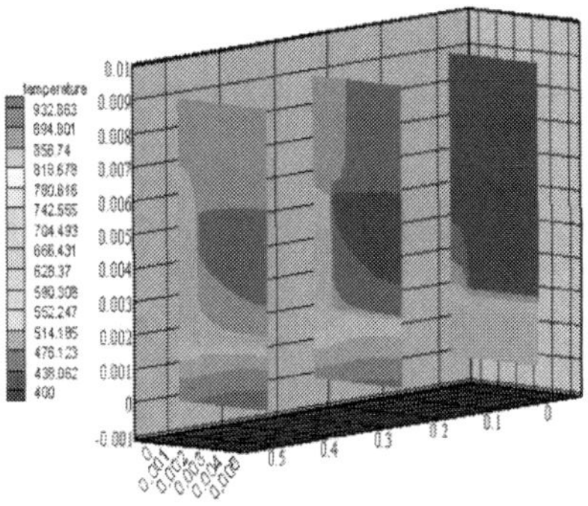

Figure 17. Cloud diagram of channel temperature distribution at inlet temperature of 400K.

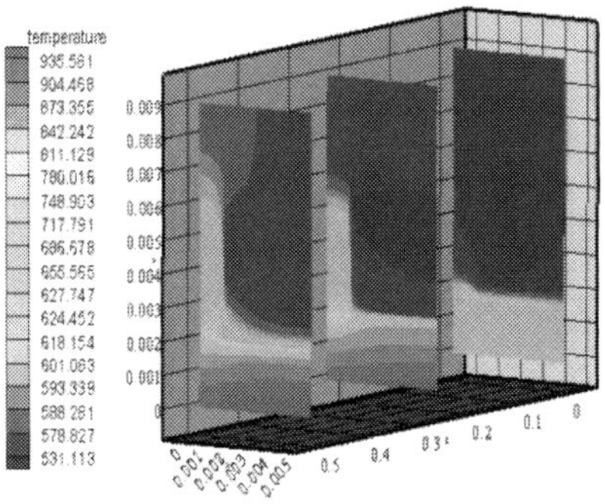

Figure 18. Cloud diagram of channel temperature distribution at inlet temperature of 500K.

4 Summary

This paper proposes a CFD simulation analysis method for regenerative cooling channels, and performs three-dimensional modeling and analysis using the current mature fluid mechanics calculation software Fluent. The temperature field of the cooling channel and the temperature distribution of the inner wall surface are given. The effects of inlet mass flow, inlet pressure, and inlet temperature on the flow and heat transfer characteristics of kerosene in the cooling channel are analyzed, and the conclusions are as follows:

(1) As the inlet mass flow increases, the maximum wall temperature and maximum oil temperature are decreasing. The larger the inlet mass flow, the lower the maximum wall temperature and oil temperature. When the inlet mass flow rate is less than 8g / s, the maximum wall temperature and oil temperature rise sharply as the mass flow rate decreases; when the inlet mass flow rate is greater than 20g / s, the maximum wall temperature and oil temperature decrease as the mass flow rate increases. Slower.

(2) The trend of wall temperature changes under different inlet pressure conditions is basically the same, all of which increase sharply, then steadily and then abruptly rise suddenly, and then decrease near the exit, which is almost unchanged, and the temperature difference under different conditions is not large. The temperature of the central flow shows a linear growth trend, and the temperature difference is not greater than 5K under different inlet conditions.

(3) The distribution trends of the centerline wall temperature, central flow oil temperature, kerosene pressure, and heat transfer coefficient along the channel are basically the same for different inlet temperatures. When the inlet temperature exceeds 200K, the centerline wall temperature of the inner wall surface changes abruptly, the wall surface temperature suddenly rises, heat transfer deterioration occurs, and then decreases, and the change near the outlet section is small. When the inlet temperature is high, the corresponding pressure is large, and the pressure loss of kerosene in the channel is also large. When the inlet temperature is low, the pressure loss is small. The inlet temperature has a greater impact on the heat transfer coefficient. As the inlet temperature decreases, the temperature difference between the wall surface and

kerosene becomes larger, and the convective heat transfer coefficient gradually increases.

References

1. Xiao Hongyu. Analysis and Design of Reinforced Active Cooling Structure Reinforcement for Super-combustion Stamping Engine [D]. Xi'an: Master's Thesis of Air Force Engineering University, 2007
2. Gong, Les. Real-Time Simulation of Aeroheating of the Hyper-X Airplane [R] NASA. 20110014692, 2011
3. X-51 A scramjet breaks record for longest hypersonic flight [R] .http: // dvicecom / archives / 2010/05 / hypersonic-x-51php, 2010.
4. Ungar, Eugene K .; Schunk, Richard G .. Spacecraft Radiator Freeze Protection Using a Regenerative Heat Exchanger [R] NASA 20110012266,2011
5. Parris, Daniel K .; Landrum, D. Brian. Effect of Tube Geometry on Regenerative Cooling Performance r [R] NASA 20050209950, 2005
6. Stephen J Scotti, Carl J Marthn. Active Cooling Design for Scramjet Engines Using Optimization Methods. AIAA 1988
7. Moholt, Matthew. ARMD Hypersonics Project Materials and Structures: Testing of Scramjet Thermal Protection System Concepts [R]. NASA. 20090012145, 2009
8. A. Frohlich, G. Scheuerer. Three-dimensional flow analysis in a rocket engine coolant channel of hogh depth / width Ratio. AIAA91-2183, 1991
9. Korabelnikov, A. V., Kuranov, A. L .. Thermal protection of hypersonic flight vehicle using chemical heat regeneration [R] .AIAA. A0214063, 2002
10. Leonard CP, Amundsen R M, Bruce III W E. Hyper-X hot structures design and coMParison with flight Data. [R] AIAA 2005-3438, 2005
11. Fan, X J; Yu, G; Li, J G; Lu, X N; Sung. Performance of Supersonic Model Combustors with Distributed Injection of Supercritical Kerosene [R]. AIAA 2007-5406, 2007

Study on Flow and Heat Transfer Characteristics of Supercritical Kerosene in a Round Tube

Feng Gao, Qian Zhang [a], Hongyu Xiao, Fengli Chen，Xuefeng Xia

Air and Missile Defense College, Air Force Engineering University, 710051 Shaanxi Xian, China

Abstract. The finite volume discrete solution of the Navier-Stokes equation and the RNG model of the turbulence model are used to numerically simulate the flow and heat transfer characteristics of supercritical kerosene in a circular tube. The results show that as the inlet mass flow increases, the wall surface temperature and the central flow oil temperature gradually decrease, and the pressure loss becomes larger. As the inlet temperature increases, the wall surface temperature and the central flow oil temperature both increase. When the heat flux density is constant, as the pressure increases, the deterioration of heat transfer will be weakened, and increasing the pressure can improve the effect of convection heat transfer.

1 Introduction

Hydrocarbon fuel, as a new type of engine fuel, has advantages that other fuels do not have, making its research very important for the development of superb technology. And studying the heat transfer flow of hydrocarbon fuel in the cooling passage of super-combustion stamping engine is helpful to understand the flow and heat transfer phenomenon and mechanism of hydrocarbon fuel in the passage. For the structure of cooling passage of super-combustion stamping engine and the entire engine, Design helps.

As the flying Mach number increases, the combustion is often in a supersonic state, which requires higher and higher fuel transportation speeds, which causes the pressure of the fuel to rise above its critical pressure, and the fuel absorbs heat in the cooling channel. Its own temperature is also rising. When it reaches the critical temperature of the pressure, it becomes a supercritical fluid. Compared with pure materials such as H2O and CO2, the flow and heat transfer of hydrocarbon fuels under supercritical pressure have similarities and have their own complex characteristics. Therefore, it is very important to study the flow and heat transfer characteristics of hydrocarbon fuel in the supercritical state.

The previous researches on inorganic working fluids and pure substances are more. However, most of the super-combustion stamping engine propellants use aviation kerosene, and there is less research on them. Kerosene is a complex mixture of various organic substances. The heat transfer flow in the channel is more complicated, and its research on the heat transfer characteristics in the cooling channel is of great significance.

This article simulates the flow and heat transfer characteristics of kerosene in a slender circular tube, analyzes the mechanism of the deterioration of heat transfer, and studies the heat transfer flow of kerosene under different inlet conditions.

2 Calculation models and numerical methods

2.1 Physical model and geometry

The research object is a two-dimensional axisymmetric round pipe with a diameter of 12.0mm and a length of 5.5m. The first section is 1.0m long without heating. It is guaranteed that the kerosene entering the heating section is a fully developed tube flow, and the heating section is 4.5 m. Figure 1 shows the calculation area and related parameters of the circular tube. The kerosene flows through the horizontal tube from left to right. Because it is only to explore the heat transfer and flow characteristics of supercritical kerosene in the tube, and the diameter of the round tube being studied is small, the effect of buoyancy is not considered in the calculation. Therefore, the numerical simulation can be simplified to an axisymmetric two-

[a] Corresponding author: zhangqian8913@163.com

dimensional plane. It greatly reduces the calculation amount and saves the calculation time [1].

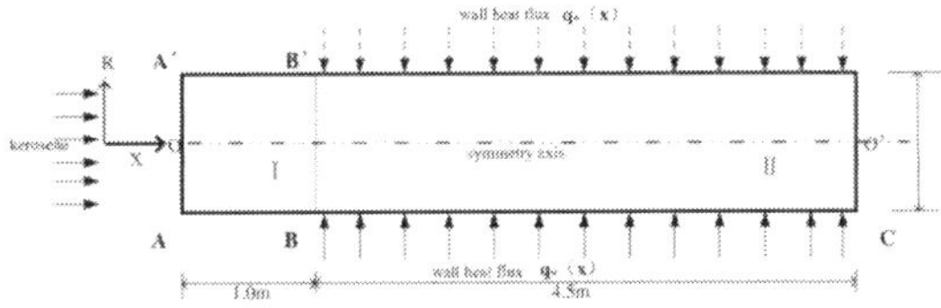

Figure 1. Physical model diagram.

2.2 Boundary conditions

The 1.0m inlet section is not heated, so that the kerosene entering the heating section can fully develop tube flow. The inlet AO is the boundary condition for the mass flow into the inlet, mass flow $\dot{m} = \dot{m}_0$, inlet pressure $P = P_0$, inlet temperature $T = T_0$, turbulent kinetic energy $k = k_0$, and turbulent dissipation rate $\varepsilon = \varepsilon_0$, where k_0 and ε_0 are shown by equations (1) and (2) respectively.

$$k_0 = \frac{3}{2}[\bar{u}(0.16Re_{D_H})^{1/8}]^2 \quad (1)$$

$$\varepsilon_0 = C_u^{3/4}\frac{k^{3/2}}{l} \quad (2)$$

Among them, C_u is the empirical constant (about 0.09); $\bar{u}$ is the average flow velocity; Re_{D_H} is the Reynolds number calculated from the hydraulic diameter; l is the turbulence length scale, and the approximate value is $0.07D$.

The wall AB is an adiabatic wall, the wall heat flux is zero, and the numerical boundary conditions are:

$$u_x = u_y = 0; q = 0; \frac{\partial p}{\partial y} = \frac{\partial k}{\partial y} = 0 \quad (3)$$

The wall surface BC is the wall surface of the heating section, the heat flux density is the heat flux density applied in the physical model, and the numerical boundary conditions are:

$$u_x = u_y = 0; q = q_w; \frac{\partial p}{\partial y} = \frac{\partial k}{\partial y} = 0 \quad (4)$$

OO 'is a symmetric boundary condition. The axial velocity, temperature, and pressure are all zero. The numerical boundary conditions are:

$$\frac{\partial u_x}{\partial y} = \frac{\partial T}{\partial y} = \frac{\partial p}{\partial y} = \frac{\partial k}{\partial y} = \frac{\partial \omega}{\partial y} = 0; u_y = 0 \quad (5)$$

The exit CO 'is a free exit, and the numerical boundary conditions are:

$$\frac{\partial u_x}{\partial x} = \frac{\partial u_y}{\partial x} = \frac{\partial T}{\partial x} = \frac{\partial p}{\partial x} = \frac{\partial k}{\partial x} = \frac{\partial \omega}{\partial x} = 0 \quad (6)$$

2.3 Governing equations

The finite volume method is used to solve the Navier-Stokes equation discretely[2-5]. The governing equation is as follows:

Mass conservation equation:

$$\frac{\partial}{\partial x}(\rho u_x) + \frac{\partial}{\partial y}(\rho u_y) = 0 \quad (7)$$

Momentum conservation equation:

$$\frac{\partial}{\partial x}(\rho u_x u_x) + \frac{\partial}{\partial y}(\rho u_y u_x) = \frac{\partial}{\partial x}(\mu\frac{\partial u_x}{\partial x}) + \frac{\partial}{\partial y}(\mu\frac{\partial u_x}{\partial y}) - \frac{\partial p}{\partial x} + S_{u_x} \quad (8)$$

$$\frac{\partial}{\partial x}(\rho u_y u_x) + \frac{\partial}{\partial y}(\rho u_y u_y) = \frac{\partial}{\partial x}(\mu\frac{\partial u_y}{\partial x}) + \frac{\partial}{\partial y}(\mu\frac{\partial u_y}{\partial y}) - \frac{\partial p}{\partial y} + S_{u_y} \quad (9)$$

Energy conservation equation:

$$\frac{\partial}{\partial x}(\rho u_x T) + \frac{\partial}{\partial y}(\rho u_y T) = \frac{\partial}{\partial x}(\frac{k}{c_p}\frac{\partial T}{\partial x}) + \frac{\partial}{\partial y}(\frac{k}{c_p}\frac{\partial T}{\partial y}) + S_T \quad (10)$$

The turbulence model uses a $RNGk-\varepsilon$ model, and its governing equation is:

$$\frac{\partial}{\partial x_i}(\rho k u_i) = \frac{\partial}{\partial x_j}\left[\alpha_k \mu_{eff}\frac{\partial k}{\partial x_j}\right] + G_k + \rho\varepsilon$$

$$\frac{\partial}{\partial x_i}(\rho \varepsilon u_i) = \frac{\partial}{\partial x_j}\left[\alpha_\varepsilon \mu_{eff}\frac{\partial \varepsilon}{\partial x_j}\right] + \frac{C_{1\varepsilon}^*}{k}G_k - C_{2\varepsilon}\rho\frac{\varepsilon^2}{k} \quad (11)$$

Among them,

$$\mu_{eff} = \mu + \mu_t$$
$$\mu_t = \rho C_\mu \frac{k^2}{\varepsilon}$$
$$C_\mu = 0.0845, \alpha_k = \alpha_\varepsilon = 1.39$$
$$C_{1\varepsilon}^* = C_{1\varepsilon} - \frac{\eta(1-\eta/\eta_0)}{1+\beta\eta^3}$$
$$C_{1\varepsilon} = 1.42, C_{2\varepsilon} = 1.68$$
$$\eta = (2E_{ij} \cdot E_{ij})^{1/2} \frac{k}{\varepsilon}$$
$$E_{ij} = \frac{1}{2}(\frac{\partial u_i}{\partial x_j} + \frac{\partial u_j}{\partial x_i})$$
$$\eta_0 = 4.377, \beta = 0.012 \qquad (12)$$

2.4 Numerical calculation method

In this paper, finite volume discrete solution is used for the Navier-Stokes equation, and the RNG $k-\varepsilon$ model is used for the turbulence model. Gambit software was used to mesh the cooling channels. In order to improve the calculation accuracy, the mesh in the near-wall region was encrypted and the encryption of the near-wall mesh was guaranteed to be in line with $y^+ \leq 1$.
The governing equation is solved according to incompressible and variable physical conditions. The speed and pressure coupling uses the SIMPLEC algorithm, and the time advancement uses the implicit Gauss-Seidel method; the convection term and the diffusivity adopt the upwind difference format. The first-order upwind accuracy is used for calculation, and the second-order accuracy is used when the convergence is fast. All variables in the governing equation have second-order accuracy in time and space. User-defined wall heat flux density and kerosene physical property parameters. The physical property parameters of kerosene in supercritical state have been given in the literature [6-10]. The kerosene at different pressures and temperatures calculated by Supertrapp developed by the National Institute of Technology and Standards (NIST) In the calculation of physical properties, the method of data fitting is used.

2.5 Model reliability verification

In order to verify the reliability of the numerical model, we need to compare the numerical calculation results with experimental data. The calculation results in this paper are compared with the experimental data given in [2] to verify the numerical method.
Figure 2 shows the comparison of the calculated wall temperature and oil temperature distribution of the tube with the experimental data. It can be seen from Figure 2 that the difference between the numerical results and the experimental data is less than 10K, which indicates that the numerical calculation can simulate the flow and heat transfer of kerosene under supercritical conditions, and has high calculation accuracy.

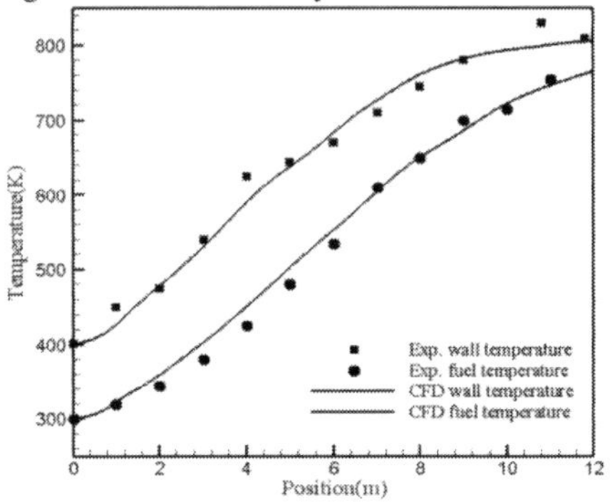

Figure 2. Comparison with experimental data.

3 Calculation results and analysis

3.1 Heat transfer characteristics at different mass flow rates

In order to study the heat transfer of aviation kerosene in a circular tube at different mass flow rates, the inlet mass flow rate was changed from 0.06 kg / s to 0.12 kg / s at an inlet pressure of 4 MPa, an inlet temperature of 300 K, and a heat flux density distribution along the axial direction as shown in Figure 3,to obtain the distribution of the inner wall surface temperature and the central flow temperature along the flow direction, as shown in FIG. 3.
Figure 3a shows the distribution of wall temperature changes under different mass flows. It can be seen that each curve shows a trend of increasing first and then decreasing. As the mass flow increases, the wall temperature gradually decreases. When each temperature curve decreases, The corresponding position in the x direction also gradually moves toward the exit direction. Figure 3b shows the change of the central flow temperature. It can be seen that the central flow temperature gradually increases with the axial direction. The larger mass flow corresponds to the lower central flow oil temperature. The observed mass flow is 0.08kg / s and 0.10kg / s. It can be seen from the two curves of that when x = 4.2m and x = 5m, the change trend of the central flow oil temperature has changed, and the temperature rise trend has become sharp. At this time, the corresponding wall temperature curve is exactly At the position just to be lowered, it can be known from the analysis in Section 3.2 that the reason for the change in the central flow oil temperature is that the heat transfer of supercritical kerosene at this time is just at the stage of heat transfer enhancement, and the wall surface transfers The heat increases, so the wall temperature starts to drop and the oil temperature starts to rise. When the mass flow rate is 0.12 kg / s, the temperature of kerosene can hardly reach the pseudo-critical temperature, so this trend is not obvious. When the mass of kerosene is 0.06kg / s, the temperature of kerosene is too high at this time, because kerosene will have obvious thermal cracking reaction at excessively high temperature (greater than 800K). The studies were performed in the temperature range below 800K.

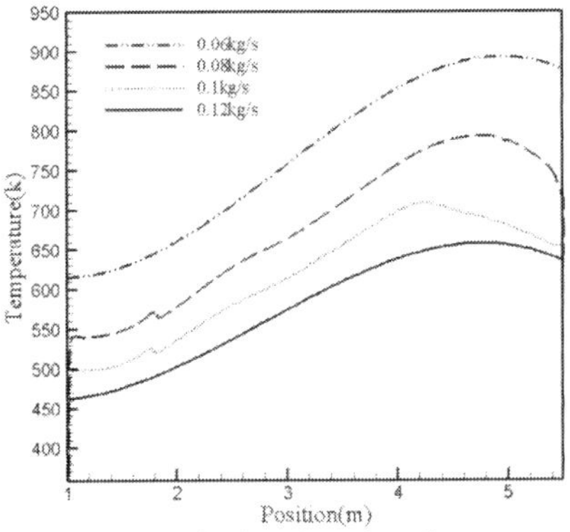

a. Wall temperature distribution along the axis

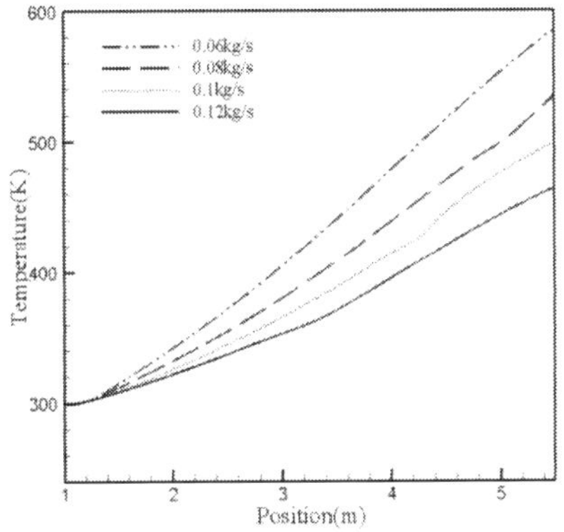

b.Center flow oil temperature distribution along the axis

Figure 3. Temperature distribution of inner wall surface and central flow oil temperature in the axial direction.

Figure 4 shows the oil pressure distribution at the central flow of different mass flow, it can be seen that the kerosene pressure along the channel gradually decreases, with the increase of the mass flow, kerosene pressure gradually increases, the corresponding pressure loss is gradually increased, which is mainly due to the increase of mass flow flow, resulting in the increase of flow rate, the flow resistance along the process increases, thus loss of kinetic energy, increased pressure loss.

Figure 5 shows the change of convective heat exchange coefficient with the wall temperature under different mass flows. As can be seen from the figure, the change trend of convection heat exchange coefficient is basically the same in the case of four mass inlets. The initial stage of heating begins to increase with the increase of temperature convection heat exchange coefficient, which is characterized by normal heat transfer phenomenon, and with the further increase of temperature, when the wall temperature reaches near the critical temperature of kerosene, the convective heat exchange coefficient begins to fall back and the heat transfer deterioration occurs, and then the heat exchange coefficient increases again, which manifests itself as the phenomenon of heat transfer reinforcement. The wall temperature change smaller in the corresponding heat transfer deterioration, both in the 650K-720K range, near the proposed critical temperature of kerosene, and the wall temperature change of secondary heat transfer reinforcement is also small, between 670K-770K, slightly higher than the proposed critical temperature of kerosene, and when the mass flow is large, the temperature of the heat transfer deterioration and reheat heat transfer is lower.

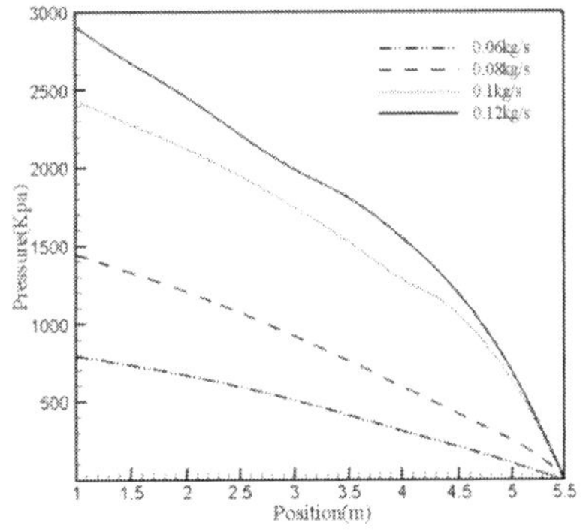

Figure 4. Oil pressure distribution at central flow

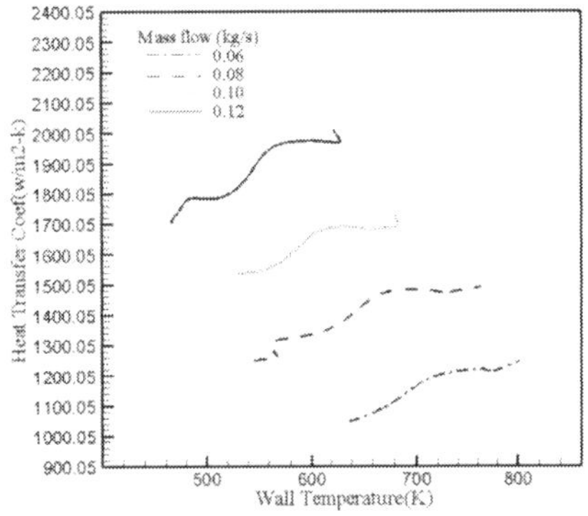

Figure5. Change of convection heat transfer coefficient at different mass flows. with wall temperature at different mass flow rates.

3.2 Heat transfer characteristics at different inlet temperatures

In order to study the heat transfer of aviation kerosene in a circular tube at different inlet temperatures, the inlet temperature was set at the inlet pressure of 4 MPa, the inlet mass flow rate of 0.1 kg / s, and the heat flux density was distributed axially as shown in Figure 3,Change from 200K to 600K. The specific entry parameters are shown in Table 1.

It can be seen from the table that as the temperature of the inlet changes, the density of kerosene and the Reynolds number of the inlet change greatly. Specifically, as the temperature increases, the density decreases, the Reynolds number increases, and the Reynolds number changes greatly, increasing from 4368. As large as 68140.8, an increase of an order of magnitude.

Table 1. Various operating conditions at different inlet temperatures

P_0(MPa)	T_0(K)	$\dot{m}$ (kg/s)	ρ_0 (kg/m^3)	Re_0
4	200	0.1	780	4368
4	300	0.1	791.56	8922.35
4	400	0.1	703.09	20906.67
4	500	0.1	611.80	39104
4	600	0.1	507.33	68140.8

Figure 6 shows the changes in wall temperature and central

 https://doi.org/10.1051/matecconf/202031603003

flow oil temperature along the axial direction at different inlet temperatures. As can be seen from the figure, both the wall surface temperature and the central flow oil temperature increase as the inlet temperature increases. When the inlet temperature is 200K, the maximum wall temperature is lower than the quasi-critical temperature of kerosene, and the temperature rise can be seen from the central flow temperature trend. However, when the inlet oil temperature is 300K or 400K, the oil temperature in the

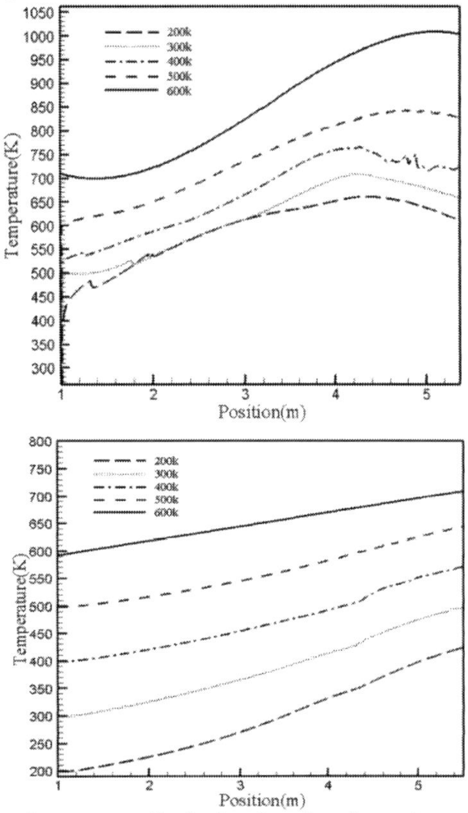

a. Wall temperature distribution along the axis **b.** Center flow oil temperature distribution along the axis

Figure 6. Axial changes of wall temperature and central flow oil temperature at different inlet temperatures.

Figure 7 shows the distribution of central flow oil pressure in the axial direction. It can be seen from the figure that the static pressure of kerosene in the pipe gradually decreases along the flow direction. Under different inlet temperature conditions, the higher the inlet temperature, the larger the central flow oil pressure The greater the flow resistance, the greater the pressure loss. This is because under the

near-wall area is critical at this time. Near the temperature, the physical properties change very drastically, and the oil temperature at the central flow also fluctuates at $x = 4\text{m}$. When the inlet temperature increases to 500K、600K, the temperature of the wall surface exceeds the critical temperature of kerosene, and the temperature of the central flow changes. And smooth

condition that the inlet mass flow rate is constant, the higher the inlet temperature is, the smaller the kerosene density is, the larger the inlet speed is, and as the flow resistance increases, the pressure loss becomes larger.Through the analysis above, we can know that when the wall temperature is higher or lower than the critical temperature of kerosene, the physical parameters of kerosene do not change much, and no heat transfer deterioration occurs. When the wall temperature is near the critical temperature, the physical parameters of kerosene in the near-wall region change drastically. The heat transfer of kerosene in a circular tube will undergo three stages: normal heat transfer, heat transfer deterioration, and heat transfer enhancement.

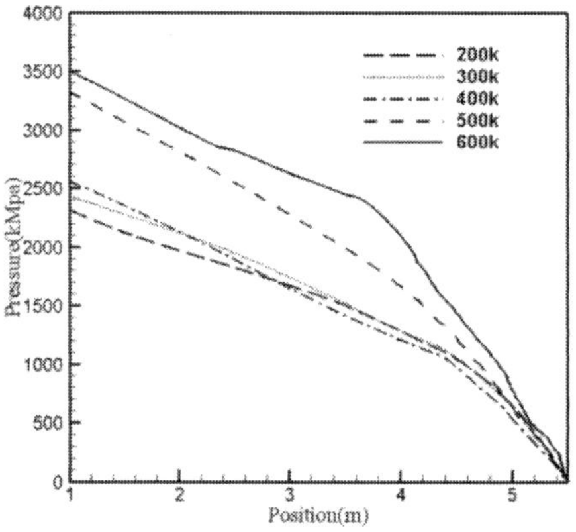

Figure 7. Variation of central flow oil pressure in the axial direction at different inlet temperatures.

3.3 Heat transfer characteristics at different inlet pressures

In order to study the heat transfer of aviation kerosene in a circular tube at different inlet pressures, the inlet pressure was changed from 2MPa to 10MPa at an inlet temperature of 300K, an inlet mass flow of 0.1kg / s, and a heat flux density of 1MW / m^2. The specific entry parameters are shown in Table 2.

Table 2. Various operating conditions at different inlet pressures

P(MPa)	$\dot{m}$(kg/s)	T (K)	Q(MW/m^2)	*Re*	ρ (kg/m^3)
2	0.1	300	1	8848.28	789.68
4	0.1	300	1	8841.94	791.56
6	0.1	300	1	8836.11	792.07
8	0.1	300	1	8835.23	794.21
10	0.1	300	1	8835.01	796.04

https://doi.org/10.1051/matecconf/202031603003

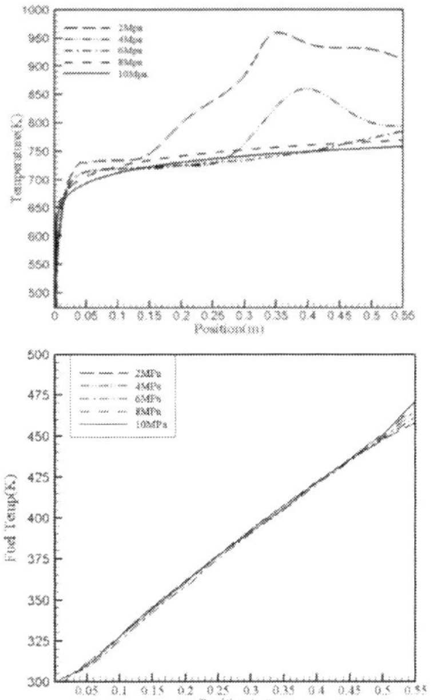

a. Wall temperature distribution along the axis **b.** Center flow oil temperature distribution along the axis

Figure 8. Axial distribution of tube wall temperature and central flow oil temperature at different pressures

Although the inlet pressure has a certain effect on the density and dynamic viscosity of kerosene, it has little effect and has a small effect on the inlet Reynolds number. Figure 8 shows the change of the wall temperature and the average oil temperature of kerosene in the tube under different pressures. It can be seen from the figure that, due to the influence of the wall heat flow density, on the one hand, the temperature gradually increases in the axial direction in the tube; on the other hand, the temperature in the near-wall region changes greatly. The results in Figure 8 (b) show that the average temperature of the kerosene in the tube basically overlaps under different pressures, and increases slightly with the increase of the pressure, which indicates that under the supercritical state of the kerosene in the tube, different pressures have little effect on the average temperature.

It can be seen from Fig. 8 (a) that the wall surface temperature changes at the inlet pressures of 2 MPa, 4 MPa and 6 MPa, but this phenomenon does not occur at high pressures of 8 MPa and 10 MPa. This is mainly due to the The sudden change in position occurred according to heat, that is, the wall temperature suddenly increased, and then gradually decreased. When there is no heat transfer deterioration, the wall temperature is basically the same, but after the heat transfer deterioration occurs, as the inlet pressure increases, the wall temperature decreases, the location of the abnormality gradually moves towards the outlet, and the wall temperature changes. The smaller the amplitude, the lower the maximum temperature, which means that the increase in pressure helps to enhance heat transfer. As the inlet pressure continues to increase, the corresponding quasi-critical temperature increases, the temperature of the fluid does not reach the quasi-critical temperature, and heat transfer degradation is not easy to occur. Figure 9 shows the temperature distribution cloud diagram of kerosene in the pipe under different pressure inlet conditions. It can also be seen from the temperature distribution cloud diagram that the temperature distribution of kerosene in the pipe is basically the same in the inlet section, but the temperature of the outlet section increases with the pressure in the outlet section. It has been increased. Comparing these five figures, it can be seen that with the increase of the inlet pressure, the heat transfer effect from the wall surface to the axis gradually increases. The kerosene flowing in the tube can take more heat from the wall surface. The increase in pressure helps to enhance the heat transfer effect.

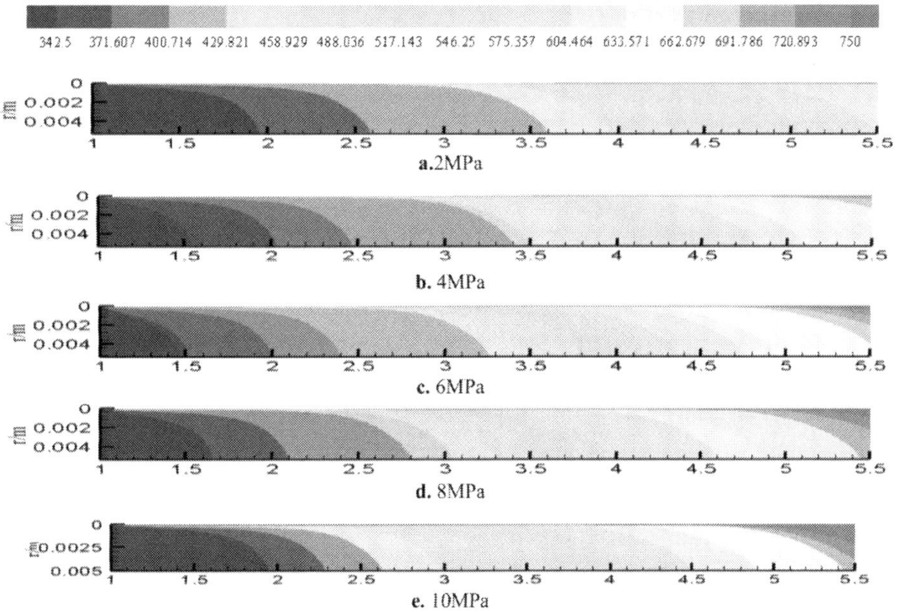

Figure 9. Temperature distribution of kerosene in the tube under different pressure inlet conditions.

4 Summary

This chapter studies the flow and heat transfer characteristics of aviation kerosene in a circular tube. Firstly, the flow and heat transfer characteristics of supercritical kerosene were studied under different heat flow density distributions, and the following conclusions were drawn:

(1) With the increase of the inlet mass flow, the wall temperature and the central flow oil temperature gradually decrease, and the pressure loss will increase. The phenomenon of heat transfer deterioration occurs near the quasi-critical temperature of kerosene. The temperature change corresponding to the heat transfer deterioration at different mass flow rates is small, all in the range of 650K-720K, which is near the quasi-critical temperature of kerosene.

(2) The inlet temperature also has a large effect on the flow and heat exchange of kerosene in the tube. The wall surface temperature and the central flow oil temperature increase with the increase of the inlet temperature. When the wall surface temperature is higher or lower than the kerosene critical temperature, The physical parameters of kerosene do not change much, and there is no deterioration in heat transfer. When the wall temperature is near the critical temperature, the physical parameters of kerosene in the near-wall region change drastically. The heat transfer of kerosene in a circular tube will undergo three stages: normal heat transfer, heat transfer deterioration, and heat transfer enhancement.

(3) Under the condition of constant heat flow density, the heat transfer characteristics of kerosene with different inlet pressure in the circular tube were analyzed. It is found that the heat transfer deterioration occurs near the quasi-critical temperature due to the drastic changes in physical properties. As the pressure increases, the heat transfer deterioration weakens, and increasing the pressure can improve the convective heat transfer effect.

References

1. Xiao Hongyu. Analysis and Design of Reinforced Active Cooling Structure Reinforcement for Super-combustion Stamping Engine [D]. Xi'an: Master's Thesis of Air Force Engineering University, 2007
2. Wang Yingjie. Review of heat transfer characteristics of supercritical fluids [J]. Science & Technology Review, 2012, **30** (33); 74-79
3. Li Xunfeng, Zhong Fengquan. Numerical study of flow and heat transfer of aviation kerosene tube under supercritical pressure. [J] Propulsion Technology. 2010.Vol.**31**.No.4
4. Xianliang Lei, Huixiong Li, Shuiqing Yu, Tingkuan Chen. Numerical investigation on the mixed convection and heat transfer of supercritical water in horizontal tubes in the large specific heat region [J]. Computers and Fluids. 2012, **64**, 127-140
5. Dong Eok Kim, Moo Hwan Kim. Two layer heat transfer model for supercritical fluid flow in a vertical tube. [J]. The Journal of Supercritical Fluids. 2011, **58** (1), 15-25
6. Fan, X J;Yu,G;Li,J G;Lu,X N;Sung. Performan -ce of Supersonic Model Combustors with Distributed Injection of Supercritical Kerosene [R]. AIAA 2007-5406, 2007
7. Gong, Les. Real-Time Simulation of Aeroheating of the Hyper-X Airplane [R] NASA. 20110014692, 2011
8. Guo Chaobang, Li Wenjie, Xing Ya. Research progress on active cooling and high temperature resistant structural components of French super-combustion stamping engines. Aeronautical Missiles, 2011 (11)
9. Ungar, Eugene K .; Schunk, Richard G: Spacecraft Radiator Freeze Protection Using a Regenerative Heat Exchanger [R] NASA 20110012266,2011
10. Moholt, Matthew. ARMD Hypersonics Project Materials and Structures: Testing of Scramjet Thermal Protection System Concepts [R]. NASA. 20090012145, 2009

Study on Model-Based System Engineering in Satellite Layout

Jiangyong Wang[a], Fa Jing, Yongquan Xia and Chunsheng Xu

Institute of Telecommunication Satellite, China Academy of Space Technology, Beijing 100094, China

Abstract. For satellite layout, it is necessary to collate manually in order to obtain accurate product technical requirements information. It is difficult to track the technical status information. It is proposed that the application of model-based system engineering (MBSE) in the cooperative development of multi-disciplinary and multisectoral satellite layout was an effective way to solve these problems. Based on the four typical processes of layout design, this paper studied the requirements to occupancy design, electromechanical joint design, integration of Attitude and Orbit Control System (AOCS) equipment layout, definition and display of equipment expansion, and give the verification design of star sensor layout. The results show that this method is beneficial to the collaboration of all departments , and improve the design ability.

1 Introduction

With the increasing demand for satellites, satellite systems are becoming more complex and need higher standards, the index has been greatly improved. Many new technologies are facing challenges. They also have the characteristics of shortening development cycle and possibly changing the technological status in the process of development, which require that the impact of the changes be given quickly. In the traditional satellite development work, a large amount of information about product technical requirements is hidden in drawings and various documents. It must be sorted out manually to obtain accurate information of product technical requirements. It is difficult to track the technical status information. In the process of satellite development, it is necessary to collect all kinds of inputs, which are controlled by human, and the workload is huge. Then, the design work is carried out according to the sequence of factors been influenced.

Demand parameters rely on manual tracking step by step. As the input for follow-up research work, Demand parameters lack traceability in the process of definition, decomposition, allocation and verification, It also lack effective digital analysis and simulation method. They often start detailed design before the scheme is mature and stable, resulting in a large number of design rework and delays. In addition, there is no complete end-to-end requirement traceability and influence analysis, resulting in many design iterations and rework. Requirements/design/validation results are mainly recorded and transmitted in the form of documents, which makes it difficult to trace and trace the whole development process, and to assess the impact of changes timely and effectively when changes occur.

In the process of mechanical and electrical joint design, although some collaborative design work has been carried out, the situation of "separating wall" between different departments/specialties is still very prominent, lack of integrated collaborative development method and carriers, a large number of format conversion when exchanging data, and duplication of work lead to poor communication and transmission of information.

Lack of early model-based virtual verification, there is no closed-loop design, mainly through the indicators of parameters to verify the validation of the benchmark (generally based on empirical data). For example, in satellite development, according to the product configuration of Attitude and Orbit Control System(AOCS) specialty, the installation requirements of sensors and actuators for satellite AOCS specialty are given, and the whole satellite layout design is carried out to meet the requirements of AOCS specialty.

Three-dimensional model has been widely used in manufacturing and production, but its role in analysis is insufficient. It is necessary to expand the application of three-dimensional model, use three-dimensional model to carry out virtual simulation, and give information on the model for other subsystems to use. Taking the three-dimensional digital satellite as the object, based on the traditional theory and method of physical sample satellite testing and analysis, the theoretical model of satellite pure digital simulation and comprehensive analysis is constructed. The reuse rate of the model is still very low, so it is necessary to reuse the model well and make the best use of the model, and provide a variety of data interfaces for other subsystems to call.

Model-based systems engineering has been applied in many fields. The National Aeronautics and Space Administration (NASA), the United States Department of

[a] Corresponding author: wangjy501@163.com

Defense and the European Space Agency are actively applying the MBSE method, and have achieved results [1, 2]. NASA has made great breakthroughs in the design of CubeSat[3][4] and FireSat[5] using MBSE method, which also reflects the advantages of MBSE method in satellite development.

NASA JPL gave a speech entitled "OpenCAE Case Study: Europa Land Concept" on INCOSE International Workshop in 2019. Based on the Europa Lander project, this paper introduced the current model-based product development environment of NASA JPL[6]. MBSE has become a research and application hotspot in the field of system engineering in recent years[7].

MBSE emphasizes the consideration of requirements, design and validation in the process of work. By establishing the relationship between validation model and requirement model, logical model, and other relevant models used in the design process, the validation coverage analysis is carried out to ensure that all projects meet the validation requirements. This paper studies the implementation of model-based engineering in satellite layout from four aspects, and takes star sensor integrated layout as an example to verify.

2 Satellite Layout and Model-based Systems Engineering

2.1 Relevant Contents of Satellite Layout

In satellite development, satellite configuration layout design is an important overall design work. The basic task of spacecraft configuration design is to combine satellite subsystems and their instruments and equipment into an internal and external space dimension coordination, to ensure the realization of spacecraft functions, and to meet the installation requirements and capabilities of instruments and equipment in each subsystem. The mechanical environment of launch vehicle is favorable to the development of spacecraft and the growth of payload capability of spacecraft. Satellite configuration is not only the main force-bearing component for designing satellite structure subsystems, but also the overall shape, volume and size allocation, overall layout and mass characteristics calculation of instruments and equipment of satellite subsystems [8].

The overall layout is an important work in satellite configuration[9]. Arrange a reasonable layout to meet the various needs of satellite missions. Layout includes: instrument and equipment layout design (including AOCS sensors); propulsion system (pipeline) layout design; cable network routing design; layout design outside the satellite; and layout analysis (sensor field of view analysis, interference analysis, mass characteristics calculation) ,and so on.

2.2 Concepts and Processes of Model-based System Engineering

INCOSE, which first proposed the concept of MBSE, gave the following definition: "Model-based system engineering is a formal modeling tool that can support activities such as requirements, design, analysis, validation and confirmation from the conceptual design stage, and continue throughout the whole development process and the subsequent life cycle stage" [10]. MBSE method takes model as the basis of system description and realizes the whole process from conceptual design, scheme design, experimental verification to engineering implementation with visualized model. Compared with document-centred system engineering method, MBSE has the advantages of high understanding and communication efficiency, easy data acquisition, good traceability of technical status, and integration of design and verification [11][12].

In general, MBSE can be cut in practical applications according to the overall process of requirements, functions, logic and physics.

1) Requirements: Describe the requirements that should be met from the perspective of system black box. Requirement model refers to the set of requirements from top to the bottom of the system, and the logical relationship between them.

2) Function: The technical function of the system proposed in response to the demand is the white box of the demand. The functional framework describes all the functions of the system and their interfacing relationships.

3) Logic: Describe the relationship of system composition, and the interfacing relationship among system, subsystem and component.

4) Physics: The Final Three-Dimensional Definition of Products.

3 Application of Model-based System Engineering in Layout Design

3.1Requirements trace back to space occupancy design

The traditional document-based requirement management mode is adopted in aerospace model development. The requirement mode has many limitations. It is difficult to change, maintain and track the change of users' requirement and its impact on the system in time, and the requirement comes from different documents. And the development of requirement management technology has solved this problem. Requirement management generally includes requirement definition, requirement allocation, requirement validation and traceability management. Traceability-oriented requirement management and control can drive and correlate requirements to downstream engineering activities.

In the design of satellite configuration layout, different requirements are defined and allocated: boundary constraints of large-scale systems (such as carry rocket), design prohibition zone, AOCS constraints, interface constraints, fixture tooling constraints, etc., and then subdivided to verify the requirements of subdivision,figure1.

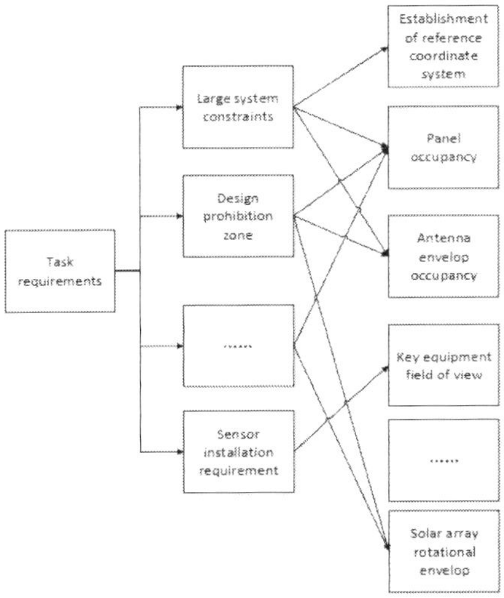

Figure 1. Construction of requirements model

According to the occupancy design, the requirements are preliminarily verified. The space occupancy of the whole satellite is determined by the whole satellite coordinate system, the structure coordinate system and the local coordinate system. The space static envelope is used to determine the space occupancy of the large parts, and the motion envelope is used to determine the moving area of the moving parts. For the key single equipment, the field of view surface is established with the constraint of the viewpoint and the main axis of the field of view. Establish the traceability pedigree of demand and occupancy design, and realize the traceability control of requirement.

3.2 Mechatronics Joint Design Based on Electrical Principle

Multi-specialty joint simulation of electrical and mechanical engineering is carried out in cable design. According to the requirements of satellite overall design, we customize and manage single or multi-level classification of electrical parts, define standard electrical connection attributes, form a unified Library of electrical components, and ensure the unification of data of electrical components in the whole professional design. Electrical schematic diagram is established to cover the whole process of system principle design. Attributes and technical parameters are automatically captured in the schematic diagram. Connectivity and quality of the schematic diagram are analyzed. Synchronization with 3D design is achieved through business rules. Improve the quality of electrical system and data continuity, make the electrical 2D-3D design process realize the integration based on database, design and optimize the accuracy of electrical 3D harness channel/cable channel in the overall product context. Through intuitive three-dimensional geometric wiring harness routing design, the structure layout of wiring harness (wiring harness structure catalogue, trunk routing, branch point setting, branch routing, etc.) is determined. By laying electrical signal wiring harness, the transmission path of electrical signal is established, and the function of electrical logic design of contact list is realized. Geometric wire harness and electrical wire harness are combined to form the whole wire harness design system, which provides information about the mass, center of mass and moment of inertia of satellite wire harness system. Through the three-dimensional annotation of electrical wiring harness, it can guide the production and assembly of follow-up workshop, and realize the design of full three-dimensional electrical system; or form flattening drawings, which are compatible with the status that some workshops need two-dimensional drawings to guide production. The real-time synchronization between the harness digital mockup and the flattening digital mockup is realized.

In view of the trend of three-dimensional geometric wire harness, the calculation rules of the bending radius of the wire harness, the design points of cable slack and the design elements of cable color are encapsulated and integrated. Table 1 is an example table of design elements, which becomes the corresponding design guide template, and is integrated with the mechanical three-dimensional design software to support mechanical design. Based on this, the staff can input the cable network design data quickly, reduce the workload of manual calculation and improve the overall efficiency of three-dimensional cable design.

Table 1. Digital cable design elements

Design activities	Elements number	design elements	DesignEss entials
Harness control	1	wire diameter	◆ …
	2	Bend radius	◆ …
	3	slack	◆ …
	4	…	◆ …

3.3 Integrated Layout Design from Control Parameters to Attitude Control Equipment

The satellite AOCS consists of attitude sensor, controller and actuator. Sensors measure satellite attitude angle and attitude angular velocity. Controller calculates control force and control moment according to the information output by sensors, and produces corresponding control effect by actuator. The layout of control system equipment is a multidisciplinary and complex issue involving control, machinery, optics and so on. Each device of the control system has a design input to the layout, including the requirements of cable distance between devices, field of view of sensors, pointing requirement of actuator (thruster), layout space of controller cabin, and precision measurement requirement of sensors,figure2.

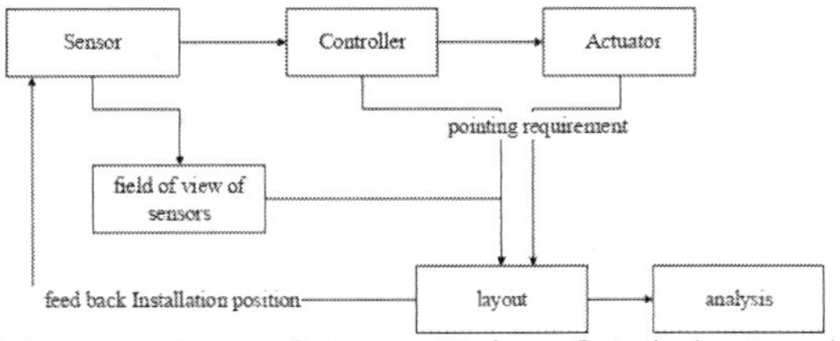

Figure 2. Integrated Layout Design of Attitude Control Equipment

Attitude control sensors and thrusters in satellites require no other parts to be shielded in their field of view, so they can be arranged accordingly. According to the occlusion results of each sensor, the effectiveness of the combined sensor to the target is obtained. Sensor field of view analysis includes field of view occlusion analysis and sensor effectiveness analysis. Field-of-view occlusion analysis can be simplified to determine whether the cone at a point in a certain direction interferes with other satellite components or other targets. A single sensor is equivalent to a point light source. Starting from the light source, the visible star surface elements are projected onto the designated polar coordinate plane to analyze the visible region of the field of view at all angles, and then calculate the occlusion of the sensor at a certain time. The field of view can be represented by a cone.

For actuators such as thrusters, the space coordinates and azimuth angles of the thruster's thrust points are placed in the layout, and the thrust and moment are calculated to determine the layout after meeting the requirements. After the layout is determined, the position and mass characteristics of the equipment layout are fed back to the control for further analysis.

3.4 Extended Definition and Integrated Display of Equipment

Satellite layout involves a large number of instruments and equipments, which are defined more extensively. Combining form data with attribute definition in three-dimensional model. The mechanical interface (including basic information, installation information, MCI information, etc.), thermal interface (basic thermal performance, temperature measurement point, heat consumption, temperature range) and remote control information include basic information of remote measurement parameters (parameter code, parameter name, parameter category). The basic information of remote control command (instruction code, instruction name, instruction type, etc.), parameter processing method, instruction criterion and channel information are all managed in a unified way. On the basis of equipment model used in layout design, the application scope of three-dimensional model is extended.

Using the model to carry out simulation analysis, we try our best to solve the problem of product assembly at the design stage, eliminate the hidden danger of downstream product manufacturing, and avoid large rework. In addition to the need for assembly design at the design stage, we also need to evaluate the assemblability of the product at the later stage of product design, as well as the simulation of the production line and launching site. Practical simulation, human-machine efficiency analysis, improve the ability of satellite disassembly simulation and human-machine simulation.

The virtual simulation is carried out by using three-dimensional model, and the information given on the model is used by other professionals to realize the automatic reading of the definition information of instrument and equipment model. Taking the three-dimensional digital satellite as the object, based on the traditional physical sample satellite test results and analysis theory, the satellite digital simulation and comprehensive analysis correction model is constructed, including three-dimensional structure analysis model, thermal performance analysis model, etc.

In order to meet the requirement of other specialties to design by using three-dimensional model, the three-dimensional model is automatically transformed into a lightweight integrated display model. The information specialty uses it to design the sequential arrangement of matrix cables and bus cables. The electrical specialty uses it to carry out voltage drop analysis, power supply and distribution simulation analysis, and the test specialty can view the distribution location, superior designers can review the layout of equipment.

4 Star Sensor Layout Verification

Verification work is carried out to control the layout design of star sensors on satellites. Star sensor is a kind of sensitive device. It is a precise instrument to measure three-axis attitude of spacecraft by stellar vector. It gives three-axis attitude parameters of coordinate system relative to inertial system in quaternion form, and completes the measurement of satellite attitude angle. According to the parameter requirements of satellite star sensor, the requirement parameters such as field of view position are sorted out, and the requirement definition and control are carried out. The space layout of the system is defined by mapping the three-dimensional model of star sensor's position requirement, and the star sensor is defined as the installation point and the field of view, and the field of view requirement is defined from the field of view,figure3. The results were input to the star-sensitive black-box model, and the black-box model was used to calculate and output various kinds of information.

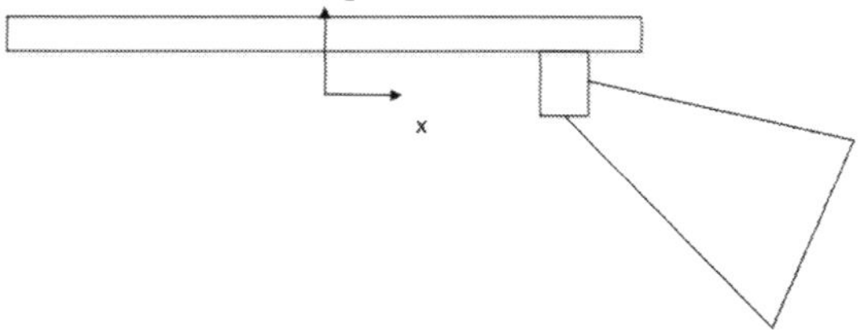

Figure 3. Star sensor Spatial Layout

5 Conclusion

Model-based system engineering has developed from exploratory research to application abroad, and there are also some research results in China[13]. This paper studies the application of model-based system engineering in satellite layout from four aspects, and validates it with the example of star sensor layout integration. It is proved that this method is of great significance to the improvement of satellite overall design level.

References

1. Kevin Vipavetz,Douglas Murphy,Samantha Infeld Model—based systems engineering pilot program at NASA Langley//AIAA SPACE 2012 Conference&Exposition.Washington D.C.:AIAA,(2012)
2. Nadia A Tepper.Exploring the use of Model-Based Systems Engineering(MBSE)to develop systems architecture in naval ship design.Cambridge:Massachusetts Institute of Technology,(2010)
3. Spangelo S C,Cutler J,Anderson L,et a1.Model based systems engineering(MBSE)applied to Radio Aurora Explorer(RAX)CubeSat mission operational scenarios//Aerospace conference.New York:IEEE,(2013)
4. Kaslow,D.Anderson,L.Asundi, S. et a1.Developing a CubeSat Model-Based System Engineering (MBSE) Reference Model - interim status. Montana:IEEE Aerospace Conference,(2015)
5. Larson w J.Applied space systems engineering.New York:McGrew—Hill Learning Solution,(2009)
6. Eric W Brower.OpenCAE Case Study:Europa Land Concept//2019 Annual INCOSE International Workshop, Torrance,CA,USA,(2019)
7. Wang Kunsheng,Yuan Jianhua,Chen Hongtao,et a1.Investigation and practice of foreign systematic engineering method based on model.Aerospace China,11(2012)
8. Peng chengrong.Overall Design of Spacecraft.Beijing:Science and technology of China press,(2011)
9. Xu fuxiang,Lin huabao,Hou shenyuan.Introduction to Satellite Engineering.Beijing:China Astronautic Publishing House,(2003)
10. HAN Fengyu,LIN Yiming,FAN Haitao.Research and Practice of Model-Based Systems Engineering in Spacecraft Development.Spacecraft Engineering, 3(2014)
11. ZHANG Youshan,YANG Lei,WANG Ping,et a1.Discussion 0n Application of Model-based Systems
12. Engineering Method to Human Spaceflight Mission. Spacecraft Engineering, 10(2014)
13. FeiFei Chen,Gang Zhao,Yong Yu,et a1. Modelica-based modeling and simulation of satellite On-orbit deployment and attitude control. Proceedings of the 2016 4th International Conference on Machinery, Materials and Information Technology Applications,(2017)

Effectiveness Evaluation of Aircraft Electromagnetic Launch System Based on RIMER

Zhiqiang Lin[1,a],Guiming Chen[1] and Hanzeng Liu[1]

[1]*Equipment management, Institute of High and New Technology, Xian, China*

Abstract. Electromagnetic launching technology is one of the important application technologies in aerospace field in the future. It will change the way of rocket launching and even be applied to aerospace kinetic energy weapon system. Aiming at the characteristics of various bottom indicators, uncertain information probability and expert participation in the effectiveness evaluation system of aircraft electromagnetic ejection system, based on the index hierarchy model, the method of RIMER(rule-base inference methodology using the evidential reasoning) is used to evaluate the effectiveness, which provides some reference for the design and development of aircraft electromagnetic launching system.

1 Introduction

Electromagnetic emission technology refers to the technology of accelerating an object to a certain speed through the Lorentz force and launching it out. With the deepening of related research, electromagnetic launching technology has been used in such fields as ship borne aircraft ejection, UAV takeoff and so on[1]. Because of many advantages of electromagnetic launching technology, it will be widely used in aerospace field, especially in rocket launching and space high-energy weapon launching. The performance evaluation of aircraft electromagnetic launching system involves many different attributes. The commonly used methods of effectiveness evaluation have their own advantages. They can get reasonable evaluation results for specific evaluation objects, but they also have limitations. The evaluation method based on belief rule base reasoning can synthesize mufti-attribute indicators with high participation of experts and good traceability and interpretability of the results. In this paper, RIMER is proposed to evaluate the performance of aircraft electromagnetic launching system, which can well solve the problems of various types of underlying indicators, uncertain information probability and requiring experts to participate in the effectiveness evaluation system of the system, and greatly improve the efficiency and correctness of the evaluation, and provide a reference for the application of Electromagnetic Launching Technology in aviation field[2].

2 Composition and performance index of aircraft electromagnetic launch system

2.1 Composition of aircraft electromagnetic launch system

The typical aircraft electromagnetic launching system includes six subsystems, as shown in Figure 1, namely, command and control subsystem, launch and control subsystem, power supply subsystem, energy storage subsystem, pulse power subsystem and electromagnetic catapult [3-4].

(1) The command and control subsystem is the command information interface between the superior command and the aircraft system, which realizes the upload and delivery of command and control information and instructions transmitted by the aircraft.

(2) The launch control subsystem is mainly composed of input of launch data, discharge control of pulse power supply, control and feedback of aircraft ejection process. Its main function is to load operational parameters of aircraft, adjust launch attitude of aircraft, control coordinated operation of subsystems, system status monitoring, etc.

(3) Power supply subsystem refers to the device that supplies power to the energy storage system by external power supply. It mainly provides stable and reliable power input to solve the energy problems needed for fast energy storage.

(4) The energy storage subsystem is an energy storage device for short time and high power electromagnetic launching of aircraft, which provides the required pulse current.

[a] Corresponding author: 727275663@qq.com

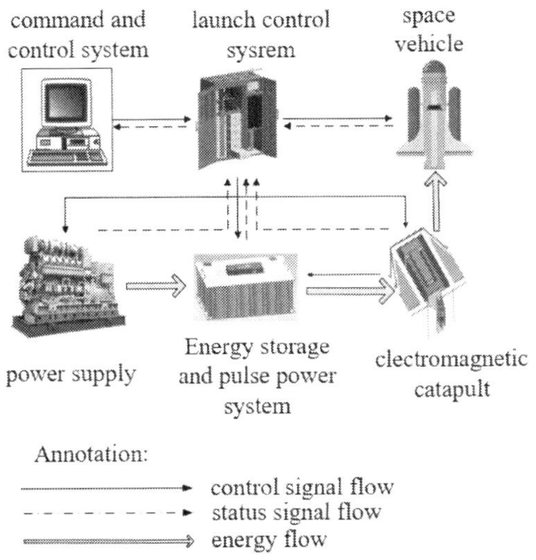

Figure 1. Composition of electromagnetic ejection system

(5) Pulse power supply subsystem mainly solves the problem of high power and fast discharge, and provides the required current for aircraft electromagnetic transmitter.

(6) Electromagnetic ejector is one of the core components of the aircraft electromagnetic ejection system. Under the control of information, the impulse current is converted into power to accelerate the aircraft, so that the aircraft can reach the required initial speed at a certain acceleration distance, and the ejection mechanism is recovered.

2.2 Hierarchical model of system performance indicators

The electromagnetic ejection system should accelerate the aircraft to a certain speed in a certain time by means of the electromagnetic acceleration device within the relative space volume and the allowable mass, and the whole process is completely controllable and stable. Therefore, we define that the performance of the system is determined by three indicators: ejection ability, environmental adaptability and controllability. Among them, ejection capability is determined by motor thrust, motor quality, motor braking capability, motor efficiency, capacity of power conversion equipment and power density of energy storage device, environmental adaptability is determined by motor maintainability, reliability of energy storage equipment and stability of control system, and controllability ability is determined by stability of control system, motor controllability, flexibility of power conversion device and so on[5]. Fault detection capability is determined. For ease of description, performance indicators are replaced by different codes.

Table 1. Performance Indicators and Symbolic Representation

CODE NAME	PERFORMANCE PARAMETER
C1	electromagnetic emission system performance
C2	ejection ability
C3	adaptive capacity to environment
C4	controllability
C5	motor thrust
C6	motor runner mass
C7	motor braking capacity
C8	motor efficiency
C9	power density of energy storage device
C10	motor maintainability
C11	reliability of energy storage device
C12	control system stability
C13	motor controllability
C14	control flexibility of power conversion device
C15	fault detection capability

Then the following hierarchical structure model is established for each capability index.

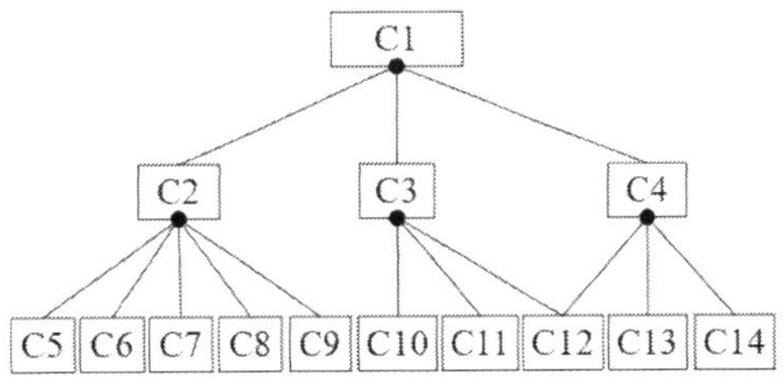

Figure 2. Hierarchical structure model of performance indicators

3 RIMER-based efficiency evaluation method

When using RIMER-based method to evaluate effectiveness, there are the following steps[6]:

1. Constructing initial confidence rule base based on experience or expert knowledge;
2. Converting input data and calculating matching degree;
3. Aggregating efficiency from bottom to top to get final effectiveness value.

3.1 Constructing an initial confidence rule base

The process is to express the capability requirement, evaluation criteria, expert knowledge and battle case experience of the relevant indicators of equipment support effectiveness in a unified format. Rule libraries are usually represented as:

$$R_k:\ if\ A_1^k \wedge A_2^k \wedge \ \ldots \wedge A_{T_k}^k,$$

$$then\ \{(r_1,\beta_{1k}),(r_2,\beta_{2k}),\ldots,(r_L,\beta_{Lk})\} \quad (1)$$

In rule k, β_{ik} denotes the reliability of the conclusion r_i when the preconditions are established.

3.2 Converting input data

Taking a special combat unit as the evaluation object, the value of the underlying capability index is collected as the input of the evaluation, and the input value is converted into the reliability structure data to calculate the matching degree of the attributes in the preconditions of the input rules. The formula is as follows:

$$v(x_i,A_{ij})=\begin{cases}\dfrac{A_{i(k+1)}-x_i}{A_{i(k+1)}-A_{ik}} & j=k(A_{ik}\le x_i\le A_{i(k+1)})\\ \dfrac{x_i-A_{ik}}{A_{i(k+1)}-A_{ik}} & j=k+1\\ 0 & j=1,2,\ldots|A_i|,j\ne k,k+1\end{cases} \quad (2)$$

Among them, x_i represents the input value and A represents the value of a certain capability index.

Calculate the matching degree α_{ij} of individual attributes for all capability indicators. The formula is as follows:

$$\alpha_{ij}=\frac{v(x_i,A_{ij})\varepsilon_i}{\sum_{|A_i|}v(x_i,A_{ij})} \quad (3)$$

Among them, α_{ij} represents the matching degree of the j value A_{ij} of input index capability requirement c_j to a single attribute.

3.3 Index effectiveness aggregation based on evidence reasoning

In this paper, the analytic algorithm is used to aggregate the effectiveness of indicators [6]. The concrete formulas are as follows:

$$\beta_l=\frac{\mu[\prod_{k=1}^{K}(v_k\beta_{lk}+1-v_k\sum_{l=1}^{N}\beta_{lk})-\prod_{k=1}^{K}(1-v_k\sum_{l=1}^{N}\beta_{lk})]}{1-\mu[\prod_{k=1}^{K}(1-v_k)]} \quad (4)$$

$$\mu=[\sum_{l=1}^{N}\prod_{k=1}^{K}(v_k\beta_{lk}+1-v_k\sum_{l=1}^{N}\beta_{lk})-(N-1)\prod_{k=1}^{K}(1-v_k\sum_{l=1}^{N}\beta_{lk})]^{-1} \quad (5)$$

Whereas β_l represents the confidence level of the l level and r represents the correction factor.

4 Performance evaluation of ejector

4.1 Establishment of rule base

According to the established performance hierarchical structure model, according to the relationship between competency factors at different levels, combined with expert opinions and experimental data, the competency of the superiors is empirically determined under different combinations of inferior competencies. Because the evaluation method of confidence reasoning will give confidence to different levels, in general, only qualitative indicators are defined as high, medium and low levels, while quantitative indicators are given three numerical levels. The rule base is established as follows:

Table 2. Rule base of C1

C2	C3	C4	satisfied	basically satisfied	can not satisfied
strong	strong	strong	0.9	0.1	0
strong	strong	weak	0.8	0.1	0.1
middle	middle	weak	0.5	0.3	0.2
strong	weak	strong	0.5	0.2	0.3
strong	weak	weak	0.3	0.3	0.4
weak	strong	strong	0.2	0.7	0.1
weak	weak	weak	0	0.1	0.9

Table 3. Rule base of C2

C5	C6	C7	C8	C9	strong	middle	weak
high	95	strong	95	high	0.9	0.1	0
high	95	strong	95	middle	0.8	0.1	0.1
high	85	middle	85	low	0.7	0.2	0.1
middle	85	weak	85	high	0.6	0.2	0.2
middle	60	weak	60	middle	0.5	0.3	0.2
middle	60	strong	60	low	0.4	0.4	0.2
middle	60	weak	95	high	0.5	0.2	0.3
low	95	weak	85	middle	0.7	0.1	0.2
low	85	strong	60	low	0.6	0.2	0.2
low	60	weak	60	low	0	0.1	0.9

Table 4. Rule base of C3

C10	C11	C12	strong	middle	weak
95	95	strong	0.9	0.1	0
95	95	middle	0.8	0.1	0.1
85	85	strong	0.7	0.2	0.1
85	85	middle	0.6	0.2	0.2
60	60	strong	0.5	0.3	0.2
60	60	middle	0.4	0.4	0.2

Table 5. Rule base of C3

C12	C13	C14	strong	middle	weak
strong	high	strong	0.9	0.1	0
strong	middle	weak	0.8	0.1	0.1
middle	low	weak	0.7	0.2	0.1
weak	high	strong	0.6	0.2	0.2
weak	middle	weak	0.5	0.3	0.2
strong	low	middle	0.4	0.4	0.2
weak	high	strong	0.5	0.2	0.3
weak	middle	weak	0.6	0.2	0.2
strong	low	weak	0	0.1	0.9

4.2 Performance value calculation

When evaluating efficiency, the value of the lowest index should be input. Three groups of different underlying performance parameters (A, B, C) are selected for evaluation.

Table 6. The value of bottom indicators

NO.	C5	C6	C7	C8	C9
A	high	98	strong	89	high
B	middle	77	weak	82	middle
C	low	55	weak	61	middle
NO.	C10	C11	C12	C13	C14
A	95	95	strong	high	strong
B	69	84	middle	middle	weak
C	73	56	weak	low	middle

Through calculation, the following evaluation results are obtained.

Table 7. Effectiveness evaluation results

equipment	satisfied	basically satisfied	can not satisfied
A	0.7329	0.1473	0.1198
B	0.6596	0.1748	0.1656
C	0.3734	0.2202	0.4046

4.3 Result analysis

From the calculation results, we can see that the overall efficiency of A equipment is better than B equipment, while C equipment is the worst, which is consistent with our judgment of the underlying indicators.

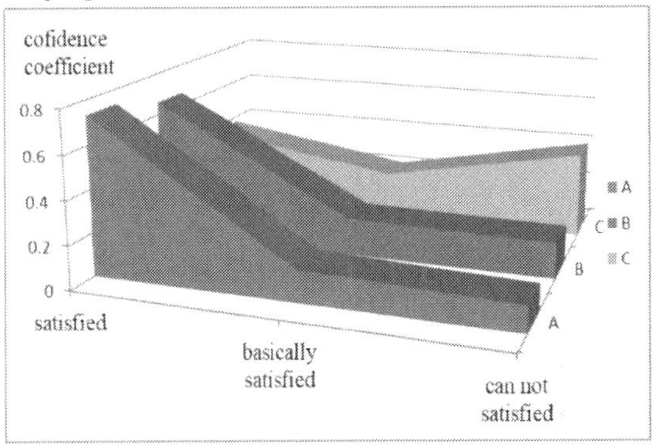

Figure 3. Schematic diagram of effectiveness evaluation results

5 Rule base optimization

Since the initial rule base is determined by empirical values, the evaluation results are not very objective. In order to obtain better evaluation results, it is necessary to optimize the confidence rule base. Based on the existing evaluation results of similar systems, this paper adopts the method of parameter optimization to deal with them [7]. The optimization model is as follows:

$$\min MSE(x, \theta, \beta) \tag{6}$$

s.t.

$$\begin{aligned} & lb_m \le x_m^k \le ub_m, \\ & k = 1,\dots,K; m = 1,\dots,M \\ & x_m^1 = lb_m \\ & x_m^K = ub_m \\ & 0 < \theta_k \le 1 \\ & 0 \le \beta_{s,k} \le 1, \quad s = 1,\cdots,S \end{aligned} \tag{7}$$

Among them, Formula (6) represents the minimum error between the result and the learning data, and Formula (7) represents the range of the precondition attribute and the initial rule weight.

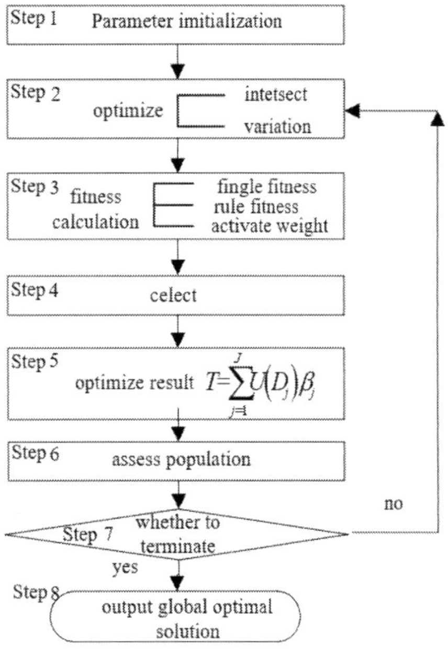

Figure 4. Flow chart of parameter optimization algorithm

The optimized rule base can be obtained by programming.

Table 8. Optimized rule base（Take C3 as an example）

weight	C11	C12	C13	C14	strong	middle	weak
0.288	60	60	60	60	0.002	0.091	0.91
0.665	64	83	91	99.555	0.728	0.274	0.01
0.028	90	75	85	67.524	0.006	0.316	0.68
0.813	61	69	100	81.102	0.324	0.392	0.28
0.209	100	100	100	100	0.946	0.028	0.04

New assessment results as follow:

Table 9. New assessment results

equipment	satisfied	basically satisfied	can not satisfied
A	0.8329	0.0973	0.0698
B	0.6096	0.2048	0.1856
C	0.2714	0.2602	0.4684

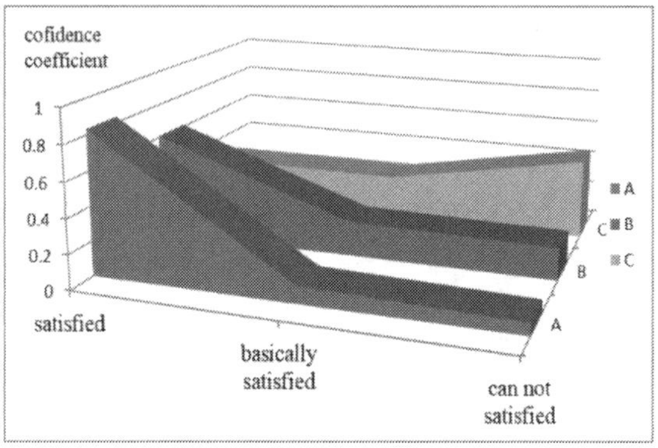

Figure 5. Schematic diagram of new results

The new evaluation results show that the gap between different equipment has widened, which is to some extent conducive to distinguishing the evaluation results. At the same time, this gap is the true reflection of the effectiveness level of each equipment.

6 Conclusion

Aerospace vehicle has high cost and high requirement of launch control, which puts forward higher requirements for the performance of aircraft electromagnetic launch system. Whether in the development and production stage or not, it is important to make a scientific evaluation of the performance of the transmitter. In this paper, the reasoning method based on confidence rule base is applied to evaluate the performance of aircraft electromagnetic ejection system. It not only eliminates the evaluation difficulties caused by qualitative and quantitative index doping, but also fully incorporates the opinions of relevant experts. At the same time, it can adjust the evaluation parameters according to the existing examples, and achieves good results. It provides a reference for the future application of Electromagnetic Launching Technology in aerospace field. It can be used for reference.

References

1. Fair H D.Progress in electromagnetic launch science and technology. IEEE transactions on Magnetics,43, 93−98,(2007)
2. Wang Y M, Yang J B, Xu D L. Environmental impact assessment using the evidential reasoning approach , European Journal of Operational Research, 174,1885-1913,(2006)
3. Mc Nab IR. Large-scale pulsed power opportunities and challenge . IEEE Transactions on Plasma Science，(2014)
4. W.A.Jacobs.Magnetic Launch Assist-NASA's Vision for the Future. IEEE. Transaction on Magnetics. (2001)
5. J.G. Baker. Electromagnetic Compatibility in Europe. IEE Proceedings Science, Measurement and Technology.(1994)

6. Chang L L, Sun J B, Jiang J. Parameter learning for the belief rule base system in the residual life probability prediction of metalized film capacitor. Knowledge-Based Systems, (2015)
7. Chang L L, Zhou Y, Jiang J, et al. Structure learning for belief rule base expert system: A comparative study. Knowledge-Based Systems,(2013)

MATEC Web of Conferences **316**, 04003 (2020) https://doi.org/10.1051/matecconf/202031604003
MEAE 2020

A C++ Based Modular Simulation Software of Satellites Earth Observation Mission Planning

Xiaoheng Liang[1], Yang Gao[2,a] and Shenggang Liu[2]

[1] *Beijing Institute of spacecraft system engineering, Beijing, China*
[2] *School of Astronautics, Beihang University, Beijing, China*

Abstract. A simulation software for efficient analysis and rapid planning of multi-satellite earth observation mission is designed to solve the problem that it is difficult to make fast decisions for multi-target and multi-constraint earth observation mission. The software is developed with C++ programming language, which has the advantages of modular design, good portability, fast running speed and less occupation of resources. Meanwhile, the input and output interfaces of the software adopt the file format of an extensible mark-up language, which can easily realize cross-platform (Windows, Linux, etc.) data exchange. The simulation results show that the software has high execution speed and precision.

1 Introduction

Earth observation plays an important role in national economic and social development (such as land census, emergency and disaster relief, etc.) and is an important part of a country's comprehensive strength [1, 2]. However, satellite earth observation and mission planning design are a very complex engineering activity [3].

Satellite mission planning is based on satellite resources to generate scheduling sequences oriented to observation tasks and satisfying various constraints. Although the number of satellites in orbit is still increasing, the limited resources of earth observation satellites are extremely precious compared with the growing observation demand. It is a more urgent problem on how to make overall planning of satellite resources and accomplish corresponding task demands with minimum resources. Mission planning of earth observation satellites plays an increasingly important role in satellite control. At the same time, the cost of most satellites has already exceeded 100 million dollars, because satellites contain a large number of complex and sophisticated technologies that are difficult to build. To make better use of satellite resources and bring them into full play in the military and national economy, it is necessary to use appropriate mathematical models and software tools to manage and distribute satellite resources and develop optimized satellite imaging plans[4].

Besides, there are many constraints on earth observation missions [5], such as sunlight, resolution, power, weather constraints, and payload type constraints. As the number of satellites, facilities and observation objects increase, the complexity of mission planning increases geometrically. Therefore, faced with a large amount of satellite, facilities and various constraints of observation targets, it is difficult for commanders to make decisions quickly according to the requirements.

Given the above problems, a number of institutions have researched the mission planning of earth observation satellites. However, most of these studies only involve single-satellite strategy (or roughly decomposing multi-satellite tasks into single-satellite tasks) and observation tasks with relatively simple constraint conditions [6], and there are few observation constraints to be considered in the task modeling process. As a result, the simulation of the real environment of the satellite and the target cannot be carried out well. Therefore, it is necessary to develop the mission planning simulation software with strong modeling ability and efficient running speed.

Multi-satellite earth observation mission planning and simulation are based on the spacecraft dynamics theory, including attitude maneuver constraints, data storage constraints, power control constraints and dozens of other constraints, to achieve the multi-satellite earth observation mission efficient analysis and rapid planning of simulation software. The simulation platform can not only be applied to the simulation of satellite earth observation mission planning and its attitude maneuver strategy but also can provide certain support for the analysis, design, and evaluation of large systems, enabling users to make decisions quickly based on the planning results. Also, for different types of observation targets (targets, area targets and trajectory targets), the planning strategy adopted by the mission planning simulation software will be optimized accordingly.

Firstly, this paper introduces the main process of mission planning simulation software. Then, the

[a] Corresponding author: 47178589@qq.com

https://doi.org/10.1051/matecconf/202031604003

composition and important functions of the software are described in detail. Finally, combined with the case test of multi-satellite to target and area target, the simulation ability of the software to the actual task is demonstrated.

2 Mission planning process

Mission planning simulation software mainly consists of configuration file loading module, unit information output module, parameter initialization module, unit access computation module, planning and scheduling module, planning result output module and task log module.

The program is developed with C++ programing language, which can easily maintain and expand. At the same time, commonly used data files are replaceable.

The input and output file format of the software is the Extensible Mark-up Language (XML). The simplicity of XML makes it easy to read and write data in any application, which quickly makes XML the only common language for data exchange [7]. Since the program is developed in C++, using XML as an input and output format means that the program can easily exchange data with Windows, Linux, and other platforms, as well as make the software portable.

The running process of the mission planning simulation software is shown in Figure 1.

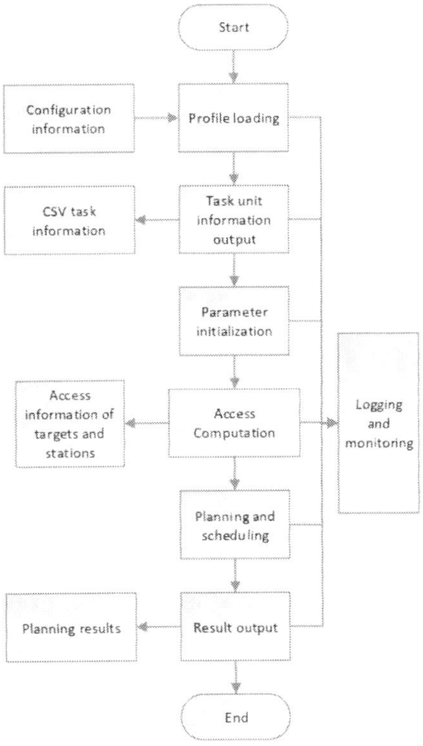

Figure 1. The execution process of mission planning

Firstly, the program loads the input XML file configured by the user in the interface format defined by the mission planning software. After the configuration file loading module completes the parsing of the input XML, the unit information output module generates a Comma Separated Values (CSV) file for user reference. Then, the parameter initialization module will establish the information matching between the satellites and the observation targets (including the facilities). The unit access computation module divides the output information of the previous module into meta-tasks and calculates the visibility of satellites to the targets and facilities. Finally, after the meta-tasks are planned and combined according to the strategy by the planning and scheduling module, the planning result output module writes the observation plans into the output XML file to finish the process.

The task log module can monitor the key steps and record the problems encountered during the whole process of mission planning.

3 Software construction and functions

3.1 Input and output

Before starting the task planner, the user needs to input requirements for the simulation task into an XML file in the mission planning interface definition format. This file contains information about the step size, total simulation duration, satellite, observation target and facilities of the mission simulation.

The partial nodes of the input XML are shown in table 1.

Table 1. Nodes of mission planning input XML

Nodes	Name	Fields
1	task attributes	clock
1.1	task interval	interval
2	satellites' info	satellites
2.1	orbit-propagator	propagator
2.2	sensors' info	sensors
2.3	satellites constrain	constrains
3	stations' attributes	stations
3.1	minimum of elevation	minElevation
4	targets' info	targets
4.1	sensors' type	sensorType
5	area targets' info	areaTargets
5.1	observation frequency	imageTimes
6	trajectory targets' info	lineTargets
6.1	observation time	imagePeriod

After the mission planning starts, the configuration file loading module first extracts the information of the input XML file by using the parser. Then, based on the extracted information, the corresponding instance is generated and stored in the memory unit to prepare for subsequent module calls. The relevant information involved in the configuration file includes mission information, satellite information, relay satellite information, earth station information and various observation target information. The execution flow of this module is shown in Figure 2.

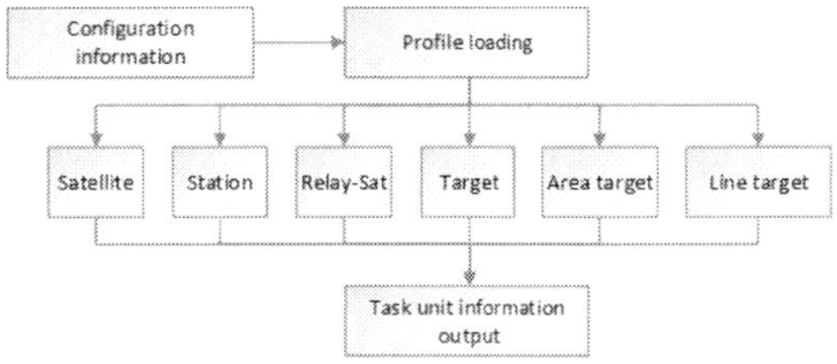

Figure 2. The process of loading configuration file

After completing the planning and scheduling of the observation task, the output module of the planning result will generate an XML file according to the output interface format of the mission planning simulation software. This file contains the task attributes, target types, load attributes, start and end time of the task, attitude maneuver angle and working time, instruction loading information and data download information of the observation task sequence. Some of the nodes that output XML are shown in table 2.

Table 2. Nodes of mission planning output XML

Nodes	Name	Fields
1	task attributes	clock
1.1	plan results	ifSuccess
2	satellites' info	satellites
2.1	plans' info	obsvPlans
2.1.1	targets' info	target
2.1.2	sensors' info	sensor
2.1.3	attitude maneuver info	attiMnv
2.1.4	commands upload info	cmdUpload
2.1.5	data download info	dataDownload

3.2 Task unit information output

The XML file has the advantage of data exchange and specification unification, but it cannot show the information of the task unit visually. The unit information output module uses input XML file information to output CSV files corresponding to task unit information according to different categories of task units (such as satellite and facilities). CSV is a common, relatively simple file format that is most widely used to transfer tabular data between programs [8]. The CSV file can display information about the same type of task unit, and the user can view the CSV file to see if the input is the same as the actual requirement.

3.3 Task initialization and modeling

After loading the input information, the whole earth observation mission needs to be modeled. First, the parameter initialization module sets the initial parameters of the task and establishes relevant constraints. For example, the total simulation time of the task, simulation step size, start and end time of the task scene, etc. Then, the module needs to forecast the orbit of the satellite and convert the coordinates of the earth station and observation target. Finally, the parameter initialization module establishes the communication matching between the satellite and the observation target, the earth station and the relay satellites.

The main flow of the parameter initialization module is shown in Figure 3.

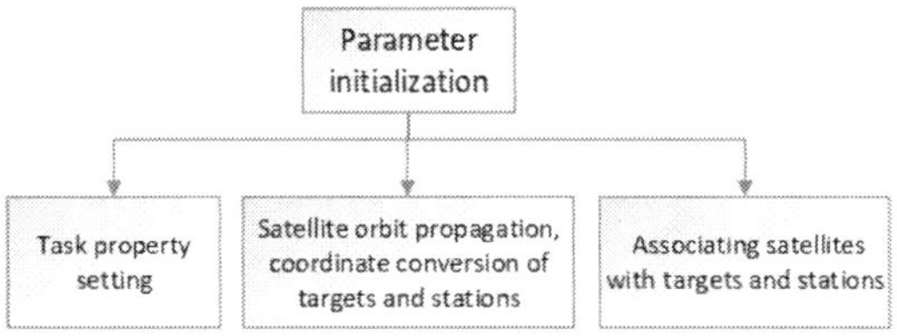

Figure 3. The initialization process of parameters

Before the simulation, some limitations/constraints need to be completed, such as the limit of the range of the satellite's attitude maneuvering ability, the minimum visible elevation of the satellite to the observed target, and the maximum number of satellites that the earth station can connect to at the same time. Establishing these constraints can improve the simulation of the real task environment.

Satellite orbit modeling is an important process of mission planning. According to the satellite information in the input XML, the orbital forecaster makes ephemeris forecast for the satellites in the mission scenario. Mission planning simulation software currently supports two-body, J2, SGP4, and High Precision Orbit Propagator (HPOP).

The description of the position parameters of targets, area targets, and general trajectory targets on the earth's surface is usually based on the earth-centered earth-fixed (ECEF) frame, but the coordinates in the earth-centered inertial (ECI) frame need to be used in subsequent access computations. This module has three coordinate conversion functions. Among them, the accuracy of the fast coordinate transformation function is similar to the conversion function provided by IERS, and the speed is much faster than the latter, which can significantly shorten the time of coordinate transformation and the whole task simulation process.

The location of the earth station is generally described by geographical coordinates, and its coordinates in the inertial reference frame are needed in the access computation. The difference between the coordinate transformation of the earth station and the observation target is that the geographical coordinates need to be converted to the geocentric reference frame coordinates first, and then to the inertial reference frame coordinates later.

In a multi-satellite to earth observation mission, each satellite may have different observation targets. In the parameter initialization module, the observation target configuration function matches each satellite and its target to be observed. In the subsequent access computation module, the program calculates only the position relationship between the satellite and the matching target. If the user limits the facilities and relay satellites that the satellite can use, the corresponding constraint information can be written in the configuration file. The module will then establish a constraint relationship between the satellite and the measurement and data transmission resources that can be invoked. If relevant constraint information is not written, the mission planning simulation software, namely the facilities and relay satellite with default input profile, can communicate with the imaging satellite.

3.4 Access computation

In earth observation, the satellite can only start to perform the observation mission when the observation target appears in the field of view of the payload (such as CCD, infrared imaging and SAR, etc.).

The unit access computation module uses the satellite forecast ephemeris generated by the parameter initialization module and the coordinates of the observation object and the measurement and control facility (earth station and relay satellites) in the inertial reference frame as input and then calls the module function to calculate the geometric relationship between the two.

The results are the visible windows (visible start and end times) between the mission elements (satellite and observation target, satellite and facility). The access computation is used for planning and scheduling module calculations, and it is output to a text file.

Before the geometric calculation, the access computation module will divide each task into units based on the time sequence, and the length of the units is based on the simulation step size of the whole task (which can be specified by the user). Targets access windows divided by time sequence are called "meta-tasks". The meta-task computing method can reduce the complexity of visibility computing and improve the efficiency of the whole mission planning process.

The access computations for satellite point-to-targets and facilities are similar, so these calculations can use the same functions. The results of the satellite to point target access computation include the start and end time of the visible arc, the attitude maneuver angle, the load type used in the observation and the resolution. The results of satellite access computations for facilities include the start and end times of the visible windows, the available duration, and the facilities (or relay satellites) number.

The visibility of the satellite to area target and general trajectory target is different from the above calculation. Before calculating the visibility of the region target and the general trajectory target, the two types of targets need to mesh, and the segmentation accuracy is specified by the configuration file entered by the user. The results of the satellite's access computation of area targets include the start and end time of the visible arc, the visible boundary, the load type used in the observation, the coverage rate and the resolution, etc.

For satellites with attitude maneuvering capability, it is necessary to increase the calculation of attitude side swing angle during the calculation of visibility (and observation target). If the satellite needs to adjust the orientation of the imaging load utilizing lateral maneuver to complete the imaging of the observation target, it is necessary to calculate whether the attitude maneuver time is within the allowed time window.

3.5 Planning and scheduling

There are three main conditions for the earth observation mission to be carried out. Firstly, the satellites should receive the loading instructions from the facilities. After that, the satellites are able to observe the specified target. Finally, data can be exchanged between satellites and designated data transmission facilities. Satellites that meet the above three conditions in time order can achieve a complete planning response.

The observation task arrangement is a complex process with repeated iterations. The program needs to calculate whether the first command loading window of each satellite and the designated observation and control station (if not specified, it needs to calculate all stations) conflicts with the first window visible to the observation target (if there is a conflict, it needs to find the closest non-conflict window). Then, it is necessary to select the nearest digital transmission window in the visible window of the observation target satisfying this condition. The data transfer window also needs to be wide enough for the satellite to upload all the generated data. If there is a conflict between before and after task scheduling, the priority of the meta-task sequence should be arranged according to the corresponding strategy. If a suitable one is not found in all the windows, the plan will be deemed ingenerated.

The multi-satellite planning includes many constrains such as optimizing the selection of the imaging orbit of the observation strip, evaluating imaging quality corresponding to the imaging time of each orbit and the time planning of the carried out observation strip. In the mission planning, it is possible to roughly estimate the duration of observation, data transmissions, the conversion time of satellite attitude and the changes of relevant on-board state parameters.

There are three different planning strategies for multi-satellite earth observation mission planning and simulation: the shortest mission response time, the best imaging quality and the highest mission completion. The main flow of the planning and scheduling module is shown in Figure 4.

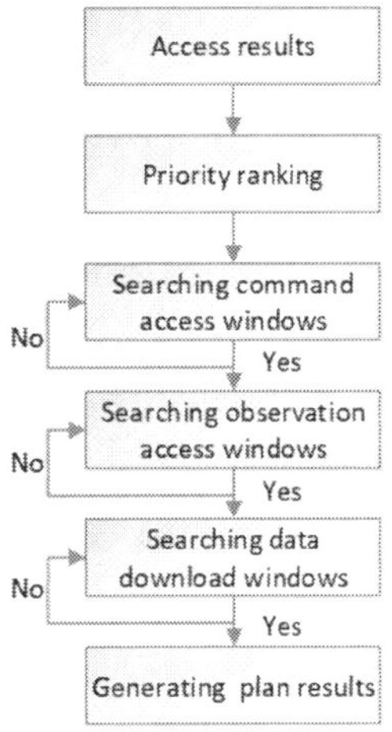

Figure 4. The process of planning and scheduling

The shortest response time strategy is the time-optimal strategy based on the first observation results of the satellite. Task requirements have different priorities for different observation targets. Therefore, the mission planning simulation software puts forward the concept of "target observation weight". The shortest response time is to make the task response time of each target as early as possible under the premise of the priority target observation weight.

The optimal imaging quality strategy is based on the resolution of the satellite observation target. The optimal imaging quality refers to giving priority to the target with "the highest imaging quality" and making the imaging resolution of each target as high as possible under the premise of the priority target's observation weight.

The strategy with the highest mission completion is the strategy with the highest number of imaging times based on the satellite observation mission frequency. The task completion rate is the highest, which means that under the premise of the observation weight of the priority target, the number of imaging times of the observed target should be as much as possible, and the task with the highest observation frequency should be first met (single-target task completion rate = the number of observation schedules of the target/the number of observation times of the target).

3.6 Logging and monitoring

Logging is one of the most important aspects of software development. It can help quickly locate problems. The task log module is used to record the exceptions during the whole mission planning simulation cycle. When the mission planning simulation software runs unexpectedly, the module will call the task log module to write the location (or possible cause) of the error to the log file.

4 Application examples

In earth observation missions, the typical objects to be observed are targets and area targets. To demonstrate the simulation capability of the mission planning software, this paper designs the targets and area targets cases respectively for test.

4.1 The targets example

7 satellites and 13 targets are used in this example. The mission time of the simulation is 1 day. The process of the observation in the simulation is shown in Figure 5. The whole running time that this simulation example takes is less than 0.75 second.

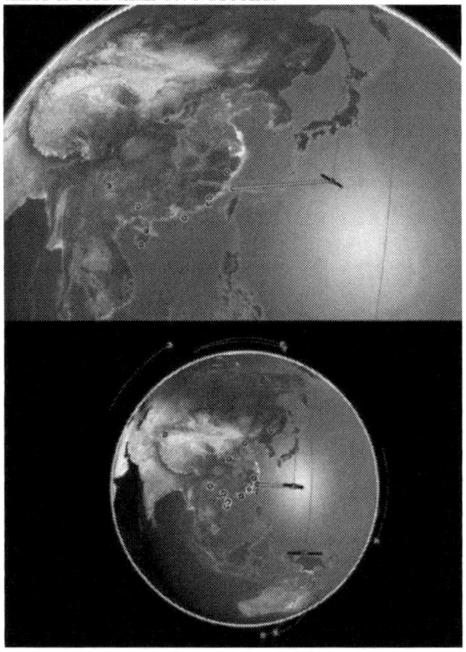

Figure 5. Multi-satellite observation for targets

4.2 The area target example

In the case of the area target, 10 satellites, an area target and 8 facilities are used as task units in this paper. The mission time of the simulation is 1 day. The process of the area target observation is shown in Figure 6. The whole running time that this simulation example takes is less than 7.03 second.

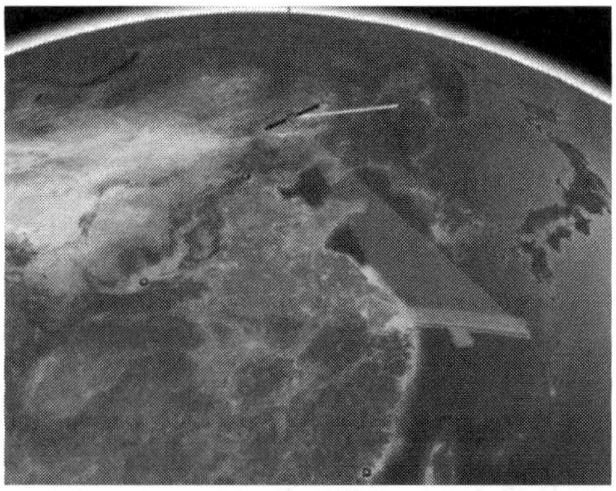

Figure 6. Multi-satellite observation for area target

5 Conclusion

To solve the problem that multi-satellites, multi-object, and multi-constraint earth observation mission planning is difficult to make a quick decision, this paper designs a modular simulation software based on C++ language.

The simulation software can apply dozens of constraints such as solar elevation angles, power supply, weather information, data transmission and storage to satellites, facilities and observation targets to simulate the real environment of many earth observation missions. The modular design makes the software easy to debug and clear in structure. The task log module can provide quick feedback for software running and troubleshooting. Adopting XML file as input and output interface format has the characteristics of standardization, user-friendliness and cross-platform, which can provide good support for data exchange. Software development makes full use of the advantages of C++ language to achieve maximum computing efficiency and resource utilization. Besides, the author continues to explore parallel computing methods, which can further improve computing speed.

References

1. Jiang Wei, Hao Hui-cheng, Li Yi-jun. Review of task scheduling research for the Earth observing satellites[J]. Systems Engineering and Electronics, 2013, **35**(9): 1878-1885.
2. Gao Hong-tao, Chen Hu, Liu Hui, et al. Development of Overseas Earth-Observing Satellite Technology[J]. Spacecraft Engineering, 2009, **18**(3):84-92.
3. Bonnet J, Gleizes M P, Kaddoum E, et al. Multi-satellite Mission Planning Using a Self-Adaptive Multi-agent System[C]// IEEE International Conference on Self-adaptive & Self-organizing Systems. 2015.
4. He Ren-jie, Gao Peng, Bai Bao-cun, et al. Models, algorithms and applications to the mission planning system of imaging satellites[J]. Systems Engineering —Theory & Practice, 2011,**31**(3):411-422.
5. Tavana M , Hatami-Marbini A . A group AHP-TOPSIS framework for human spaceflight mission planning at NASA[J]. Expert Systems with Applications, 2011, **38**(11):0-0.
6. Sun Kai, Xing Li-ning, Chen Ying-wu. Agile earth observing satellites mission scheduling based on decomposition optimization algorithm[J]. Computer Integrated Manufacturing Systems, 2013, **19**(1):127-136.
7. Calvanese D, Giacomo G D, Lenzerini M. Representing and Reasoning on XML Documents: A Description Logic Approach.[J]. Journal of Logic & Computation, 2018, **9**(3):295-318.
8. Wang N, Parthasarathy S, Tan K L, et al. CSV:visualizing and mining cohesive subgraphs[C]// Acm Sigmod International Conference on Management of Data. 2008.

Research on the Application of Carbon Foam Absorbing Materials in Wireless Testing of Spacecraft Thermal Vacuum Test

DU Chun-lin[a], YIN Xiao-fang, QIN Jia-yong, LIU Xiao-ning, SUN Yu-wei, FENG-Yao

Beijing Institute of Space Environment Engineering, Beijing,100094

Abstract: With the rise of mobile communication satellites and low-orbit Internet satellite constellations, wireless testing is required in spacecraft thermal vacuum test, and a comprehensive thermal testing environment for vacuum low-temperature absorption is needed. This paper clarifies the key parameters that need to be paid attention to in the application of vacuum absorbing materials under vacuum and low temperature environment, gives reference indexes, and puts forward the technical process and test results for the development of absorbing shroud.

Keywords: carbon foam; thermal test; wireless test; absorbing shroud

1 Preface

The traditional vacuum thermal test of domestic spacecraft uses wired testing, that is, the satellite payload is connected to high-frequency or low-frequency cables, which is led out of the container through a sealed vacuum flange, and connected to ground testing equipment to complete the test. With the development of satellite antenna technology and the rise of low-orbit Internet satellite constellations in recent years, the development process proposes wireless testing during the vacuum thermal test of satellites to achieve test coverage and effectiveness. Because of this demand, it is necessary to establish a absorbing environment in a vacuum low temperature environment, and the absorbing environment must also cover the frequency band required for satellite testing.

This paper analyzes the key parameters that need to be paid attention to when the absorbing material is used in the vacuum low temperature environment, proposes the corresponding reference index requirements, and introduces the use of the absorbing material based on the work that has been carried out.

2 Introduction to absorbing materials

The basic principle of absorbing materials is to convert electromagnetic waves into other forms of energy such as thermal energy, electrical energy or mechanical energy and consume them. It finds and uses the radar stealth technology. As the core part of radar stealth technology, the absorbing material has become a hotspot in the field of military materials because of its relatively simple technology and relatively low design difficulty[1] [2] [3]. In addition to the military stealth field, absorbing materials are also key materials in the fields of defense military and civil technology, such as microwave darkroom, electromagnetic information leakage protection, electromagnetic radiation protection, building absorbing materials, etc. [4]

The absorbing material is generally composed of a matrix material (or a binder) and an absorbing medium (absorbent). The binder is the base material to ensure the mechanical properties of the material; the absorbent is the loss medium, which is the key component that determines the absorption performance of the absorbing material. The performance, quantity and matching selection of the absorbing material are the main links in the design of the absorbing material[5]. There are many classification methods of absorbing materials. According to their loss mechanism, they can be divided into dielectric loss type absorbing materials and magnetic loss type absorbing materials. According to their functional configuration, they can be divided into coating type and structural type [6].

3 Application technology process

3.1 Determination of performance parameters

The absorbing materials used in radar stealth technology must meet the requirements of "thin, light, wide, strong" and corrosion resistance [1]. The ideal absorbing material should have the characteristics of absorption frequency bandwidth, light weight, thin thickness, good physical and mechanical properties, easy to use, etc. [7] [8] However, in the vacuum low temperature environment, whether the absorbing material for wireless testing of spacecraft needs to be thin, wide, light and strong, the analysis is as follows:

[a] Corresponding author: to_cl2004@126.com

① The structure of the current wave absorbing material is a pyramid structure, which belongs to the structure type rather than the coating type. The wave absorbing material as the test tool does not contact the test product, so it can have a certain height;

② satellites generally involve antennas of different frequency bands and bandwidths, and the frequency bands of different satellites are not the same, so the absorbing material only needs to meet the reflectivity requirements of the corresponding frequency band of the satellite, so the "wide" frequency band may not be used as a key parameter;

③ During the test, the absorbing material is static and can be temperature controlled, so the structural strength is not high;

④ The absorbing material is generally installed through a metal frame. Light weight can reduce the degeneration of the metal frame under the alternating temperature environment, and thus maintain the stability of the absorbing device's absorbing performance. Therefore, "light" can be used as a key parameter.

Based on the above analysis, it is known that reflectance and density are the key parameters that need to be considered when the absorbing material is used in a vacuum and low temperature environment. In addition, environmental adaptability should also be considered: such as condensable volatiles, water vapor reabsorption rate, surface emissivity, Vacuum quality loss and other parameters. According to experience, the key parameter reference indexes of recommended absorbing materials under vacuum and low temperature environment are shown in Table 1.

Table 1 Reference parameters of key parameters of absorbing materials

Key parameters	Reference indicators
Reflectivity	less than -25dB
Density	less than 3g / cm³
Condensable Volatile Matter (CVCM)	less than 0.1%
Moisture Vapor Absorption Rate	Less than 1%
Surface emissivity	greater than 0.85
Vacuum quality loss (TML)	less than 2%

3.2 Performance testing and selection of absorbing materials

When selecting the absorbing material, the indicators of performance parameters need to be tested by authoritative organizations. Based on previous literature research and performance test results, two materials, carbon foam and silicon carbide, were initially selected; each material has two structural forms, open-cell foam and closed-cell foam. At the same time, the mesh structure of the foam material is also different. Therefore, the selection of materials needs to be based on the test environment. In this project, considering the low-temperature vacuum environment, the materials cannot be closed, so the open-cell foam material is selected, and the cost is also considered. Therefore, this project chooses Open-cell carbon foam materials are used as research objects. Carbon foam materials (hereinafter referred to as "carbon foam") are structural and dielectric loss-type absorbing materials.

The key parameters of carbon foam materials are as follows: the corresponding frequency reflectance is -28dB, the density is less than 0.5g / cm³, the condensable volatiles are less than 0.01%, the water vapor reabsorption is 0.85%, the surface emissivity is 0.896, and the vacuum quality loss is 1.39 %. In addition, the thermal conductivity of the material is also a parameter that needs attention. The thermal conductivity of carbon foam materials is 0.45W / K · m.

3.3 Development and test of absorbing shroud

After the materials are selected, the structural form of the absorbing material is determined, and the assembly process is cured. The principle prototype of the absorbing device is developed for testing the absorbing performance in a vacuum low temperature environment. If the verification test fails, you need to improve the principle prototype and test again. After the test meets the requirements, develop the absorbing device required for the formal test and carry out the formal test.

4 Carbon foam structure shaping

An open-cell carbon foam absorber is shown in Figure 1. Carbon foam is prepared by preform molding and high-temperature pyrolysis process. The pyrolysis process is used to control the electrical conductivity of the carbon foam, thereby regulating the electromagnetic parameters. Carbon foam has excellent broadband frequency response characteristics, and its 1-18GHz broadband electromagnetic parameters are shown in Figure 2.

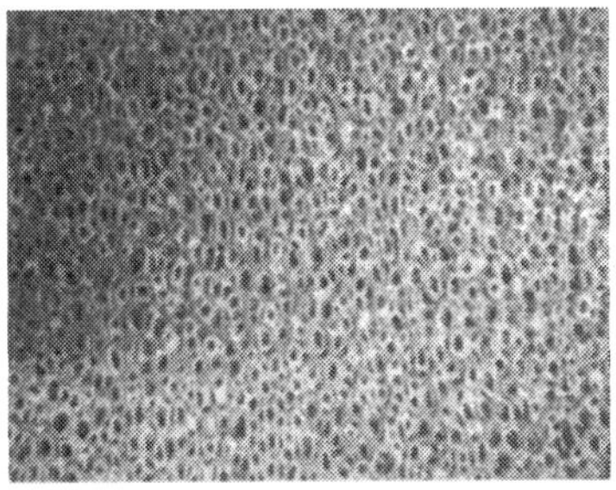

Figure 1. Open-cell carbon foam absorbing material

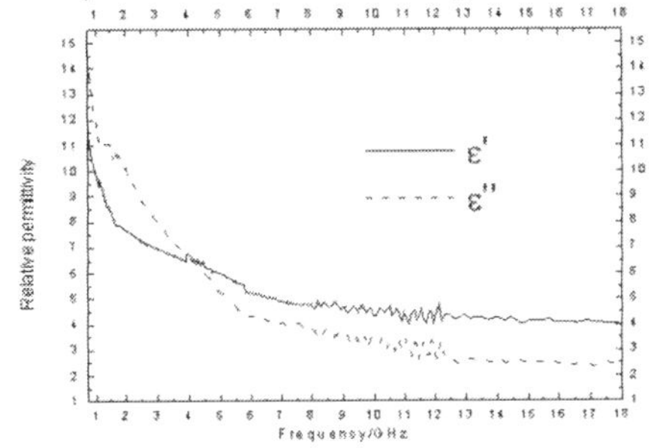

Figure 2. Broadband electromagnetic parameters of open-cell carbon foam

https://doi.org/10.1051/matecconf/202031604004

According to Maxwell's equation, two basic parameters, that is, the dielectric parameter ε and the permeability μ, can be used to characterize the electromagnetic properties of materials. These two parameters are complex numbers. Carbon foam is a dielectric loss-type absorbing material, so it is important to evaluate the dielectric parameter ε. Under alternating electromagnetic field conditions:

$$\varepsilon = \varepsilon' - i\varepsilon'' \quad (1)$$

In the formula: i is an imaginary unit, and the relative dielectric constant εr is generally used to characterize the properties of materials in engineering.

$$\varepsilon_r = \frac{\varepsilon}{\varepsilon_0} = \varepsilon_r' - i\varepsilon_r'' \quad (2)$$

Where ε0 is the dielectric constant parameter value of the material in free space.

The real and imaginary parts of the dielectric constant have specific physical meanings: the real part can characterize the degree of polarization of a material under the action of an electric field, and the imaginary part is a measure of the energy dissipated when a material is polarized in an electric field; The electromagnetic parameters of wave materials are often closely related to the composition, structure size, and micro-morphology of the wave absorbing material. Absorbing materials have contradictory relations in terms of increasing electromagnetic losses and meeting impedance matching conditions[8]. Based on the above electromagnetic parameters, the electromagnetic simulation design of carbon foam was carried out, and the specific dimensions of the integrated combined carbon foam pyramid were determined, as shown in Figure 3: flat carbon foam A with a thickness of 14mm; and the size of the bottom side of the sharp carbon foam B is 50mm×50mm, the size of top side is 5mm×5mm and 70mm high.

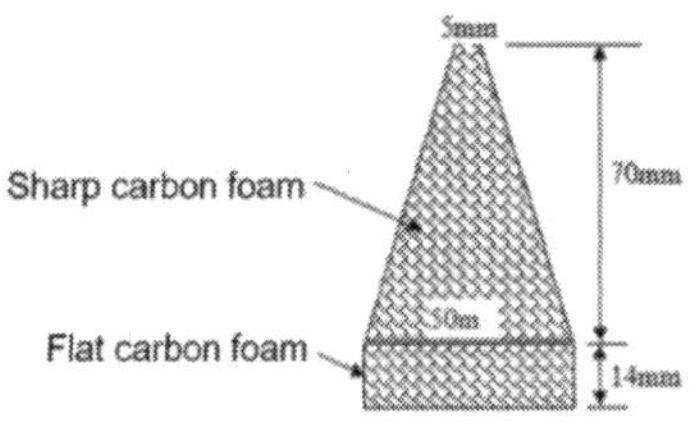

Figure 3. Simulation model of carbon foam composite structure and specific dimensions

The actual measurement results of the carbon foam absorbing performance provided by the third-party identification department are shown in Figure 4. Both the simulation and the actual measurement results show that the carbon foam absorbing performance is better than -28dB in the test required frequency band.

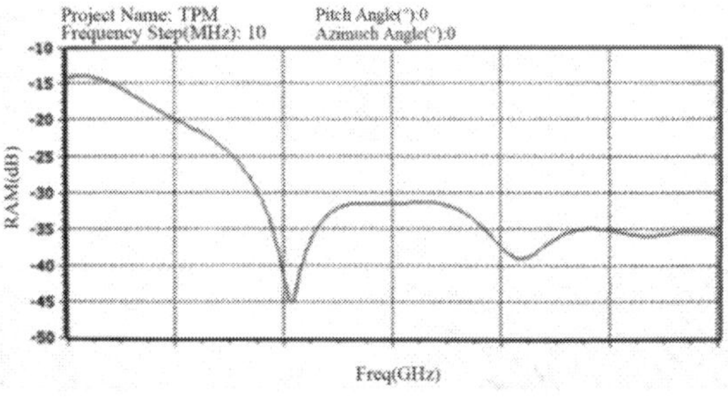

Figure 4. Third-party test results of the integrated carbon foam pyramidal wave-absorbing performance

5 Test

The shaped carbon foam pyramid was used to develop a heat-absorbing shroud, as shown in Figure 5. During the test, the absorbing shroud is placed in a container, and the two feed antennas are fixed inside the absorbing device with a non-metal plate. The high-frequency cable leads out of the cabin and connects to the test equipment, as shown in Figure 6. [9]

Figure 5. Test state of absorbing shroud

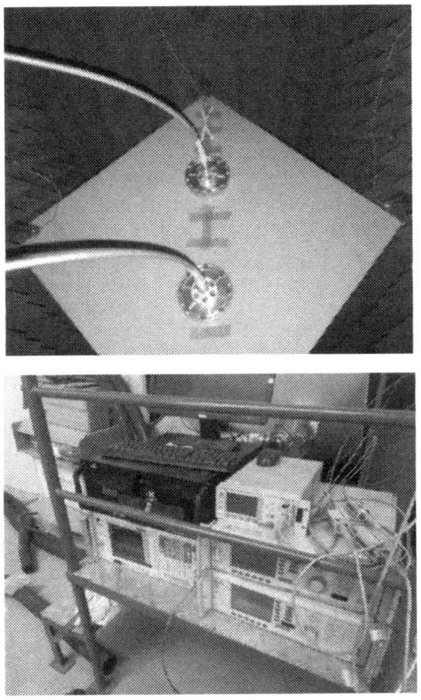

Figure 6. Test status and test equipment of the feed

The application requirements of this time absorbing material are: the reflectance within the specified frequency range of the test is ≤ -25dB, and the passive intermodulation (PIM) level indicator is less than -150dBm in the required power mode.

The test results in a low-temperature vacuum environment show that the carbon foam pyramid and the installation process can meet the requirements for use in a vacuum environment. The temperature of the internal pyramid surface is controllable, and the results of the temperature control of the absorbing shroud are shown in Figures 7 and 8. It can be seen that the internal and external temperature trends of the wave absorbing shroud are basically the same. At the high temperature equilibrium stage, the internal and external temperature difference is about 20℃(high inside).

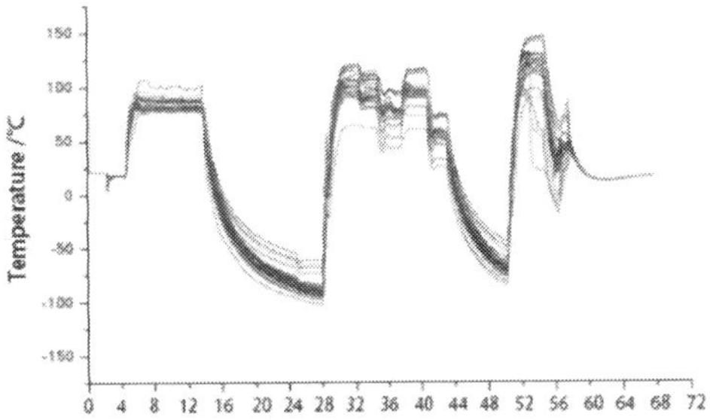

Figure 7. Temperature control results of absorbing shroud

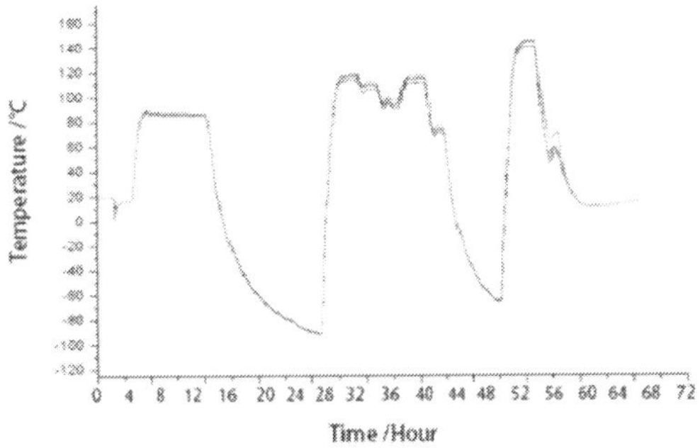

Figure 8. Temperature data of the external temperature measurement point of the wave absorbing shroud

The absorbing shroud was tested in a vacuum and low temperature environment for 5 days. During this period, the absorbing index PIM level test device has been turned on for testing. The test results are shown in Figure 9. It can be found in the range of -60 ℃~ + 90 ℃ The PIM level is better than -150dBM. When the temperature of the absorbing box reaches 100 ℃ (the internal cavity temperature is about 115 ℃ ~ 119 ℃), a sudden peak of the absorbing index appears, reaching a maximum of -138dBM for 4 seconds. It did not reappear after that, and it did not affect the test conclusion.

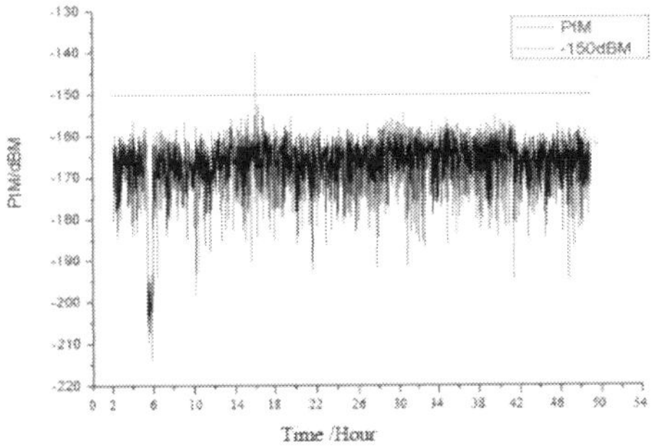

Figure 9. PIM level of the absorption shroud during the test

6 Conclusion

Absorbing materials are mostly porous materials and may contain organic components. These are factors that are not conducive to use in a vacuum and low temperature environment. Based on many years of experience, this paper gives the key parameters and reference indicators that need to be considered when using the absorbing material in a vacuum and low temperature environment. The development process and actual test results can provide references for subsequent use. It must be noted that wireless testing of spacecraft in a vacuum and low temperature environment requires different concerns. For example, if good thermal control uniformity is required, the thermal conductivity of the absorbing material should be paid attention to; Must strictly control the assembly process; the use of pyramidal absorber structure also needs to consider broken cones and excess issues.

References

1. Luo Fa, Zhou Wancheng, Jiao Yan, et al. Research status of high temperature absorbing materials [J], Aerospace Materials Technology, 2002, (1), 8-11;
2. F. Moglie, D. Micheli, S. Laurenzi, M. Marchetti and V. P. Mariani, "Electromagnetic Shielding Performance of Carbon Foams," Carbon, Vol. 50, No. **5**, 2012, pp. 1972-1980. doi: 10.1016/j.carbon.2011.12.053;
3. D. Micheli, C. Apollo, R. Pastore and M. Marchetti, "XBand Microwave Characterization of Carbon-Based Nanocomposite Material, Absorption Capability Comparison and RAS Design Simulation," Composites Science and Technology, Elsevier, Vol. 70, No. **2**, 2010, pp. 400-409;
4. Hu Xiaosai, Shen Yong, Wang Liming, et al.Study on the structure, properties and application of microwave absorbing materials[J], Applied Chemical Industry, 2015, **44** (9): 1741-1746;
5. Wang Zupeng, Yu Mingxun, Pan Bingbing, et al. Development in study on fiber absorbing materials [J], New Chemical Materials, 2010,**38** (4): 13-15;
6. Cui Zhiping, Li Xiaobo, Research Progress of Absorbing Materials [J], China Chemical Industry and Trade, 2011, **44** (1): 1-5;
7. Pang Jianfeng, Ma Xijun, Xie Xingyong, Research progress of electromagnetic absorbing materials [J], Electronic Components and Materials, 2015, **34** (2): 7-12;
8. Hou Yongshen, Wang Jianjiang, Xu Baocai, et al. Application Status and Research Progress on Wave Absorbing Materials Based on Computer Aided Design Technology [J], Materials Review A: Overview 2014, **28** (1): 127-131;
9. DU Chun-lin, QIN Jia-yong, YIN Xiao-fang, et al. Research on Microwave load Wireless PIM test of spacecraft in thermal vacuum environment [J]. Spacecraft Environment Engineering, 2017,**34** (4): 370-375.

An Analytical Method for Coaxial Helicopter Ground Resonance

Xiangyi Liu[1,a], Shuyan Liu[1], Jing Wu[1], Shansong Song[1] and Guocai Hu[2]

[1]*School of Basic Sciences for Aviation, Naval Aviation University, Yantai, China*
[2]*Department of Aviation Technology, Nanshan Aeronautical College, Yantai, China*

Abstract. A time-frequency analytical method is presented to analyze physical mechanism of coaxial helicopter ground resonance. Eigenvalue calculation and numerical integration of disturbance equations of motions are used to obtain modal characters and time-domain response characters of coaxial helicopter ground resonance, and the interaction between rotors and body is revealed according to response of various DOFs. The analysis results show that regressive lag mode with upper rotor character is the most instability mode. In dynamic instability region, coaxial helicopter ground resonance is mainly due to energy transferred between periodic lag motion of upper rotor and body roll rotation. For this instability mode, energy transferred between periodic lag motion of lower rotor and body roll rotation is also existed, and it can enhance ground resonance instability of coaxial helicopter.

1 Introduction

Helicopter ground resonance is one of dynamics problems must be solved in model design. For a single rotor helicopter with tail, the domestic and foreign scholars have done a lot of research about ground resonance, from Investigation analysis for ground resonance[1,2], the influence of the design parameters[3-5], nonlinear analysis of the helicopter ground resonance[6-8], active or passive control[9,10] and ground resonance test verification[11,12]. Many well-known conclusions have been made and the problem of ground resonance in model development has been well solved[13-16]. Coaxial helicopter compared with single rotor helicopter, the most significant difference is that body is equipped with reverse rotation of upper/lower rotor. In the published literature, there are few papers on physical investigation of coaxial helicopter ground resonance. We find that there is no systematic analysis and explanation about the influence of physical investigation and design parameters of upper/lower rotor in ground resonance, therefore, this poses potential uncertainties in the design analysis、use and maintenance of coaxial helicopter.

Referencing to [2], rotor periodic lag motion、body pitch motion and body roll motion is expressed as rotation DOFs in complex coordinates, aiming at the analytical model for single rotor helicopter ground resonance without structural damping, and physical investigation of single rotor helicopter ground resonance is revealed according to the intrinsic relation between complex eigenvector and rotor-fuselage mutual stimulate, and applies this method to analysis for a helicopter ground resonance[17]. Coaxial helicopter has two reverse rotation rotors (upper rotor and lower rotor), adding the impact of rotor and fuselage structure damping, so the relationship of various modes of eigenvectors become complex, and increasing physical analytical difficulty[18]. Because the damping that endangers the most unstable mode of helicopter ground resonance is minimal, and numerical integration of disturbance equations of motions are used to obtain modal response, at the same time it is not difficult to obtain process data of speed and acceleration of various DOFs, then we can analyze energy transfer relationship of various DOFs. Therefore, this paper presents a time-frequency analytical method for the coaxial articulated rotor helicopter to analyze the interaction between lag (upper rotor and lower rotor) and body motion, and revealing the physical investigation of coaxial helicopter ground resonance.

2 Analytical model

Each rotor(upper rotor and lower rotor) of coaxial helicopter has N blades, with the reverse rotation speed of Ω, as shown in Figure 1. When looking down, the lower rotor rotates counterclockwise, the hub coordinate system of lower rotor is $o_1x_1y_1z_1$, and following the hub movement, the origin o_1 is located at the center of the hub of lower rotor, o_1x_1 points to the tail of the fuselage; Upper rotor rotates clockwise as viewed from above, Upper rotor hub coordinate system is $o_2x_2y_2z_2$, and the origin o_2 is located at the center of the hub of upper rotor, o_2x_2 also points to the tail of the fuselage; All lag hinge extension is e, and lag angles of blades are ζ_1、ζ_2, azimuth angles of the blades are ψ_1 and ψ_2, respectively; The fixed coordinate system is $oxyz$, which coincides with the body coordinate system at rest. Assuming that the center of hub and body rotate around the

[a]Corresponding author: 281726288@qq.com

instantaneous center, roll angle around x-axis is φ_x, the instantaneous center is located at h_x below the center of hub of lower rotor, and elevation angle around y-axis is φ_y, and the instantaneous center is located at h_y below center of hub of lower rotor, the distance between upper rotor and lower rotor is h. For convenience, regarding the instantaneous center as the origin of coordinates, For the model of helicopter testing in wind tunnel, the instantaneous center is located at the universal hinge off uselage (adjust the center of mass and universal hinge coincidence), this time, h_x=h_y.

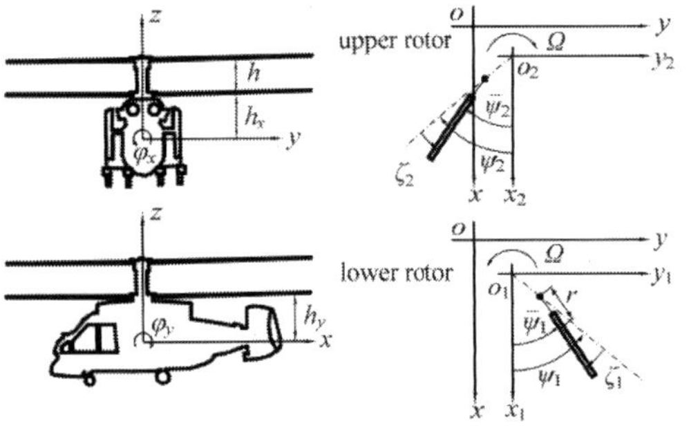

Figure 1. Analytical model of ground resonance for coaxial articulated rotor helicopter

Upper rotor and lower rotor are connected to the power transmission system via respective rotor shaft, and connected with body through the transmission system shell. Thus, there are coupling effects between upper rotor and lower rotor, as well as rotor and power/transmission system, from the torsional vibration system consisting of helicopter rotor/power/transmission. In the case of helicopter ground resonance, periodic lag motion (upper rotor and lower rotor) is coupled to motion of the centre of hub, however, there is no direct coupling between upper rotor and lower rotor.

According to Figure 1, considering small angle, the distance coordinate between blade of lower rotor and profile of lag hinge r in the fixed coordinate system is:

$$\begin{cases} x_1 = e\cos\psi_1 + r\cos\bar{\psi}_1 + h_y\varphi_y \\ y_1 = e\sin\psi_1 + r\sin\bar{\psi}_1 - h_x\varphi_x \end{cases} \tag{1}$$

Where: $\bar{\psi}_1 = \psi_1 - \zeta_1$.

Assuming that elastic deformation of upper rotor shaft is not taken into account, the coordinate of arbitrary profile of upper rotor blade is:

$$\begin{cases} x_2 = e\cos\psi_2 + r\cos\bar{\psi}_2 + (h+h_y)\varphi_y \\ y_2 = -e\sin\psi_2 - r\sin\bar{\psi}_2 - (h+h_x)\varphi_x \end{cases} \tag{2}$$

Where: $\bar{\psi}_2 = \psi_2 - \zeta_2$.

Finding the first and second derivative of time about formula (1) and (2), and obtaining speed and acceleration of blade arbitrary profile. Supposing lag stiffness of upper rotor and lower rotor blade is K_ζ, and structural damping is C_ζ, then lag motion equation of the k-th blade of lower rotor(for the sake of simplicity, the subscript k is omitted) is:

$$\begin{aligned} &\int_0^{R-e} \bar{m}\ddot{x}_1 \sin\bar{\psi}_1 r\mathrm{d}r - \int_0^{R-e} \bar{m}\ddot{y}_1 \cos\bar{\psi}_1 r\mathrm{d}r \\ &+C_\zeta\dot{\zeta}_1 + K_\zeta\zeta_1 = 0 \end{aligned} \tag{3}$$

Equation of upper rotor blade lag motion is:

$$\begin{aligned} &\int_0^{R-e} \bar{m}\ddot{x}_2 \sin\bar{\psi}_2 r\mathrm{d}r \\ &+\int_0^{R-e} \bar{m}\ddot{y}_2 \cos\bar{\psi}_2 r\mathrm{d}r + C_\zeta\dot{\zeta}_2 + K_\zeta\zeta_2 = 0 \end{aligned} \tag{4}$$

Where: $\bar{m}$ is density of blade mass; R is rotor radius. Considering small lag angle, firstly, expanding the formula (3) and (4), and then obtaining disturbance equation of k-th blade lag motion(upper rotor and lower rotor) after omitting the small above second order and second order, The disturbance equations of each blade lag motion can be converted to the disturbance equations of rotor periodic lag motion by using multi-blade coordinate transformation. For ground or hovering helicopter, if number of rotor blades are three or more, periodic coefficient with time caused by azimuth angle of blade can be eliminated after multi-blade coordinate transformation, and the disturbance equation of constant coefficient of rotor periodic lag motion is obtained. In this way, the disturbance equations of lag motion of lower rotor after multi-blade coordinate transformation are：

$$\begin{aligned} &I_b\ddot{\zeta}_{1s} + S_b h_y\ddot{\varphi}_y + C_\zeta\dot{\zeta}_{1s} - 2\Omega I_b\dot{\zeta}_{1c} \\ &+[K_\zeta + \Omega^2(eS_b - I_b)]\zeta_{1s} - \Omega C_\zeta\zeta_{1c} = 0 \end{aligned} \tag{5}$$

$$\begin{aligned} &I_b\ddot{\zeta}_{1c} + S_b h_x\ddot{\varphi}_x + C_\zeta\dot{\zeta}_{1c} + 2\Omega I_b\dot{\zeta}_{1s} \\ &+[K_\zeta + \Omega^2(eS_b - I_b)]\zeta_{1c} + \Omega C_\zeta\zeta_{1s} = 0 \end{aligned} \tag{6}$$

The disturbance equations of lag motion of upper rotor after multi-blade coordinate transformation are:

$$\begin{aligned} &I_b\ddot{\zeta}_{2s} + S_b(h+h_y)\ddot{\varphi}_y + C_\zeta\dot{\zeta}_{2s} - 2\Omega I_b\dot{\zeta}_{2c} \\ &+[K_\zeta + \Omega^2(eS_b - I_b)]\zeta_{2s} - \Omega C_\zeta\zeta_{2c} = 0 \end{aligned} \tag{7}$$

$$\begin{aligned} &I_b\ddot{\zeta}_{2c} - S_b(h+h_x)\ddot{\varphi}_x + C_\zeta\dot{\zeta}_{2c} + 2\Omega I_b\dot{\zeta}_{2s} \\ &+[K_\zeta + \Omega^2(eS_b - I_b)]\zeta_{2c} + \Omega C_\zeta\zeta_{2s} = 0 \end{aligned} \tag{8}$$

Where: cosinoidal and sinusoidal components of periodic lag angle of lower rotor are ζ_{1c} and ζ_{1s}; cosinoidal and sinusoidal components of periodic lag angle of upper rotor are ζ_{2c} and ζ_{2s}; blade moment of inertia around lag hinge is I_b; blade static moment around lag hinge is S_b.

Setting linear damping and stiffness of body around the x-axis are C_x and K_x , respectively, and damping and stiffness around the y-axis are C_y and K_y, thus, the motion equations of body pitch and roll are:

$$\begin{aligned} &\sum^N \int_0^{R-e} h_x\ddot{y}_1\bar{m}\mathrm{d}r + \sum^N \int_0^{R-e} (h+h_x)\ddot{y}_2\bar{m}\mathrm{d}r \\ &= I_x\ddot{\varphi}_x + C_x\dot{\varphi}_x + K_x\varphi_x \end{aligned} \tag{9}$$

$$\begin{aligned} &-(\sum^N \int_0^{R-e} h_y\ddot{x}_1\bar{m}\mathrm{d}r + \sum^N \int_0^{R-e} (h+h_y)\ddot{x}_2\bar{m}\mathrm{d}r) \\ &= I_y\ddot{\varphi}_y + C_y\dot{\varphi}_y + K_y\varphi_y \end{aligned} \tag{10}$$

The motional equations after expanding and multi-blade coordinate transformation are:

$$\frac{1}{2}NS_b h_x \ddot{\zeta}_{1c} - \frac{1}{2}NS_b(h+h_x)\ddot{\zeta}_{2c} + I_x\ddot{\varphi}_x + C_x\dot{\varphi}_x + K_x\varphi_x = 0 \quad (11)$$

$$\frac{1}{2}NS_b h_y \ddot{\zeta}_{1s} + \frac{1}{2}NS_b(h+h_y)\ddot{\zeta}_{2s} + I_y\ddot{\varphi}_y + C_y\dot{\varphi}_y + K_y\varphi_y = 0 \quad (12)$$

Where: the body inertial moment around the x-axis and y-axis are I_x、I_y; fuselageinertialmoment around the x-axis and y-axis are I_{fx}、I_{fy}.

The formula (5)~(8) are lag equations of motions of upper rotor and lower rotor, formula(11) and (12) are equations of body motions, and they combine to form the coupling equations of motions between rotor and fuselage of coaxial helicopter .

Setting $\mathbf{q} = [\zeta_{1s} \quad \zeta_{1c} \quad \zeta_{2s} \quad \zeta_{2c} \quad \varphi_x \quad \varphi_y]^T$,the systematic equations of motions can be written in the form of matrix:

$$\mathbf{M}\ddot{\mathbf{q}} + \mathbf{C}\dot{\mathbf{q}} + \mathbf{K}\mathbf{q} = \mathbf{0} \quad (13)$$

Where: systematic mass matrix, damping matrix and stiffness matrix are **M**、**C** and **K**.

Then, ordering $\mathbf{Q} = [\dot{\mathbf{q}} \quad \mathbf{q}]^T$, the systematic equations of motions can be rewritten as the equations of state:

$$\dot{\mathbf{Q}} = \mathbf{A}\mathbf{Q} \quad (14)$$

Where: the state matrix is $\mathbf{A} = \begin{bmatrix} -\mathbf{M}^{-1}\mathbf{C} & -\mathbf{M}^{-1}\mathbf{K} \\ \mathbf{I} & \mathbf{0} \end{bmatrix}$.

For formula (14), the characteristic analysis can be carried out, and initial disturbance response can also be calculated, if two methods can be combined, it will show a more complete systematic dynamics characteristic.

3 Analytical method

3.1 Simulation parameters

In the case of an articulated rotor, the parameters used are similar to the model used by the NASA Ames Research Centre[13], except for type of rotor, blade, and body structure damping, as shown in Table 1.

Table 1 Parameters of rotor and body

Parameter	Unit	Value
Number of rotor blade		3
Rotor radius	m	0.8108
Bladechord	m	0.0419
Offset of lag hinge	m	0.0851
Inertia moment of blade about hinge	kg.m^2	0.0173
Distance of blade centre to hub centre	m	0.2429
Mass of blade (outside lag hinge)	kg	0.2432
Damping ratio of lag	%	10.4
Distance of body rotation centre to lower	m	0.2405
Distance of upper rotor to lower rotor	m	0.2405
Inertia moment of pitch rotation	kg.m^2	0.607
Natural frequency of pitch rotation	Hz	1.59
Spring constant of pitch rotation	kg.m/rad	6.186
Damping ratio of pitch rotation	%	32.0
Inertia moment of roll rotation	kg.m^2	0.177
Natural frequency of roll rotation	Hz	3.96
Spring constant of roll rotation	kg.m/rad	11.201
Damping ratio of roll rotation	%	18.58

3.2 The results of eigenvalue and analysis

According to the calculation results of eigenvalues, modal frequency and modal damping are plotted along with rotor speed. The results of the basic model are shown in Figure 2(a) and Figure 2(b). LA1 and LA2 express advance lag mode with significant lower rotor and upper rotor characteristic respective, and LR1 and LR2 express regressive lag mode with significant lower rotor and upper rotor characteristics, BX and BY, respectively, show body roll mode and body pitch mode. If the upper rotor turns positive, then in the fixed coordinate system, LA2 and LR2 modes follow the rotation of upper rotor, and their frequencies are positive Since the rotary direction of lower rotor is opposite to that of upper rotor, therefore, in the fixed coordinate system, the rotary directions of LA1 and LR1 modes should be the same with lower rotor, and their frequencies in the fixed coordinate system should be negative. However, in order to facilitate the comparison of modal frequencies, and the frequencies of LA1 and LR1 modes are displayed as positive.

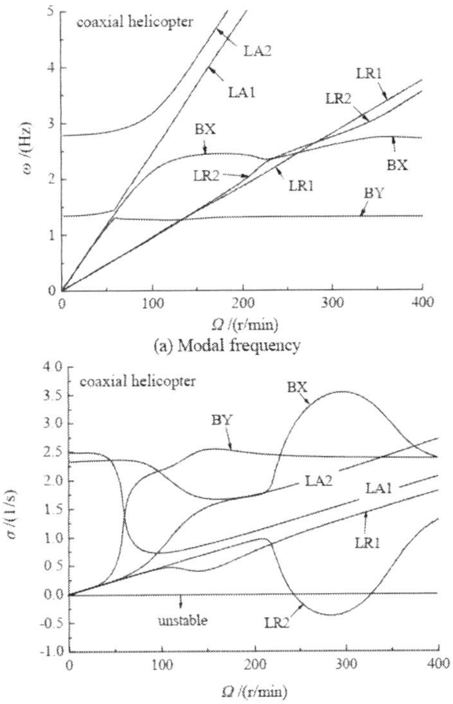

Figure 2. Modal frequency / damping vs rotor speed

As can be seen from Figure 2(a), LA2、LR2 modes and body roll mode have strong coupling. LR2 mode is mainly regressive lag mode of upper rotor, and fundamental lag natural frequency of modal rotor blades is: $\omega_\zeta=(eS_b/I_b)^{1/2}\Omega=0.4345\Omega$. The base frequency of isolated rotor regressive lag mode is: $\omega_{LR}=\Omega-\omega_\zeta=0.5655\Omega$(upper rotor and lower are same),Therefore, its rotary direction is the same as that of upper rotor (clockwise) and is coupled to body roll mode of advance rotation (clockwise) in the fixed coordinate system. LR1 mode performs regressive lag mode characteristic with upper rotor, its rotary direction is the same as that of lower rotor (counterclockwise), and LR1 mode is not coupled to fuselage mode of advance rotation. From modal frequency curves in Figure 2(a), it is found that modal frequency changes almost linearly along with rotor speed. When rotor speed is lower than unstable center (near 280 r/min), LR2 modal frequency is higher than LR1 modal frequency. otherwise, they are just opposite. It shows that body roll motion provides a different "spring" effect on LR2 mode in the coupling region of body roll mode. The frequency of the body roll mode also corresponding changes in the coupling region, which indicates that periodic lag motion of upper rotor provides different "spring" effect on body roll mode.

From modal damping curves in Figure 2(b), modal damping of LR2 is greatly reduced in the speed region where regressive lag mode is coupled with body roll mode, and modal damping is less than zero in a certain rotary speed zone, however, body roll modal damping is improved greatly. LR1 modal damping is almost unaffected by body, indicating that there is no coupling between LR1 mode and fuselage. Therefore, LR2 mode is coupled with body roll mode, and leading to lower LR2 modal damping, it is the main reason to produce coaxial helicopter ground resonance.

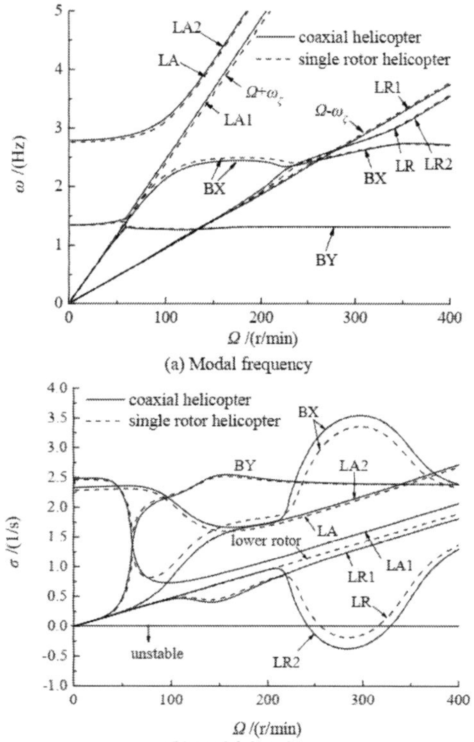

(a) Modal frequency

(b) Modal damping

Figure 3. Comparison of helicopter ground resonance modal characters between coaxial rotor and single rotor (body inertia moment is constant)

In order to understand the effect of lower rotor on coaxial helicopter ground resonance, lower rotor is retained, but inertial coupling between rotor and fuselage is removed, that is, inertial coupling terms of motional equations of lower rotor lag and body pitch and body roll are removed, and obtaining motional equations of single rotor helicopter ground resonance with only upper rotor and fuselage coupled, but body inertial moment remains unchanging, so the virtual configuration of single rotor helicopter and coaxial helicopter can be compared. modal frequency and damping curves of the single rotor helicopter ground resonance are shown in Figure 3(a), Figure 3(b). In order to facilitate comparison, curves of coaxial helicopter ground resonance are also shown in the same Figures. In the figure, solid lines indicate coaxial helicopter, and broken lines indicate single rotor helicopter.

Due to the removal of inertial coupling effect between lower rotor and fuselage, in fact, lower rotor behaves as lag characteristic of isolated rotor, from frequency curves in Figure 3(a), it is found that lag frequency of coaxial helicopter lower rotor is very close to it. It can also be seen from the curves that coupling effect of coaxial helicopter LR2 mode and body roll mode is stronger than that of single rotor helicopter. it is found that LR2 modal damping of coaxial helicopter is significantly lower than that of single rotor helicopter from modal damping curves in Figure 3(b). So it is shown that lower rotor enhance ground resonance instability of coaxial helicopter.

3.3Rotors-fuselage interaction analysis

Numerical integration of disturbance equations of motions are used to obtain the change of response of various DOFs along with time. Taking initial angle as $\zeta_{1s}(0)=\zeta_{1c}(0)=\zeta_{2s}(0)=\zeta_{2c}(0)=0.01$, $\varphi_x(0)=\varphi_y(0)=0$, and initial angular velocity is zero, rotor speed is Ω=280 r/min. From curves in Figure 3, it is found that LR2 modal damping is less than zero and other modal damping is greater than 1.0. Therefore, except for LR2 mode, other modes are attenuated in a very short time in the time course of response of various DOFs. In order to minimize impact of other modes, it needs to take response after a sufficiently long integration time for analysis. it shows response curves of coaxial helicopter various DOFs in Figure 4(a), (b), And Figure 5 shows response curves of single rotor helicopter various DOFs.

Comparing the response of coaxial helicopter various DOFs and single rotor helicopter various DOFs, what we can see is that phase relationship between periodic lag motion of upper rotor and body roll motion of coaxial helicopter is almost the same with that of single rotor helicopter , but in the same initial conditions, its response is much greater than that of single rotor helicopter after the same 8 second, it is Indicated that LR2 modal damping of coaxial helicopter is less than LR modal damping of single rotor helicopter. And amplitude of periodic lag response of coaxial helicopter lower rotor is about half that of upper rotor.

We can analyze phase relationship of various DOFs from the response curves. For LR2 mode in the fixed coordinate system, the periodic lag vector will be rotated at the frequency of LR2 mode by following rotary direction of upper rotor (or single rotor),and upper rotor(or single rotor) lag sinusoidal component ζ_{2s}(or ζ_s) and body roll vector φ_x are close to opposite phase(for an undamping system, they are just in opposite phase),while cosinoidal component ζ_{2c} (or ζ_c) is ahead of sinusoidal component about 90 °, and the corresponding angular velocity vector is ahead of angular displacement vector about 90°; On the other hand, upper rotor lag sinusoidal component ζ_{1s} and body roll vectorφ_x are close to same phase(for an undamping system, they are just in same phase), while cosinoidal component$\zeta_{1c}$is ahead of sinusoidal component about 90°, also, and angular velocity vector is ahead of angle displacement vector about 90°, It is shown that both rotor periodic lag motion and body roll motion are rotated with LR2 modal frequency ω_{LR2} following rotary direction of upper rotor. it is found that body roll motion and body pitch motion with are same phase from curves in Figure 4, Figure 5, but amplitude of roll motion is much larger, which is consistent with phenomenon of helicopter ground resonance when body is anisotropy. In order to facilitate analysis, interaction between body pitch motion and rotor is not considered, and interaction between body roll motion and rotor periodic lag is mainly analyzed. The phase relationships of various DOFs of coaxial and single rotor helicopter are shown in Figure 6(a) and Figure 6(b), respectively.

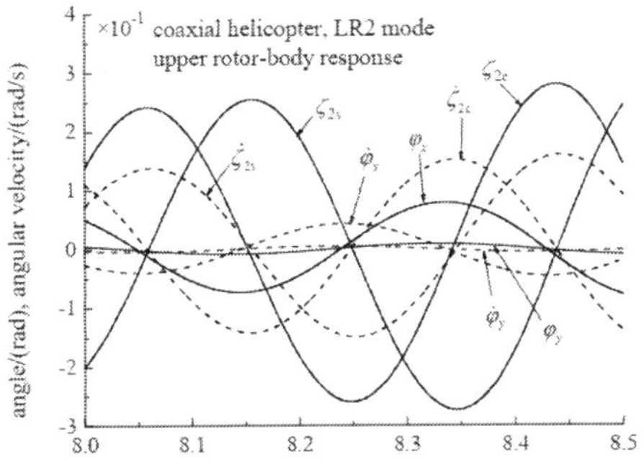

(a) Upper rotor and body

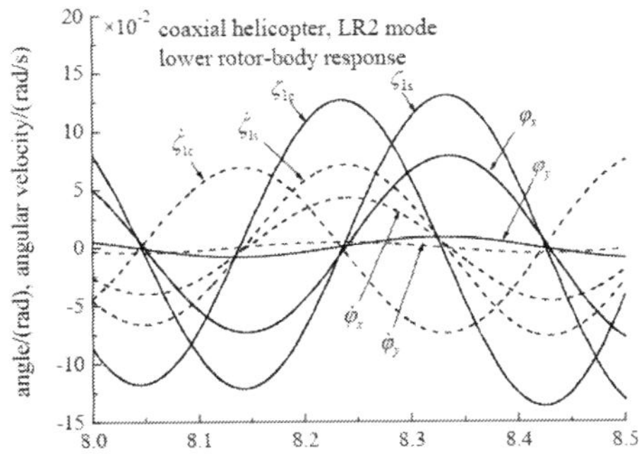

(b) Lower rotor and body

Figure 4. Response of various DOFs(Coaxial rotor helicopter, LR2 Mode, Ω=280r/min)

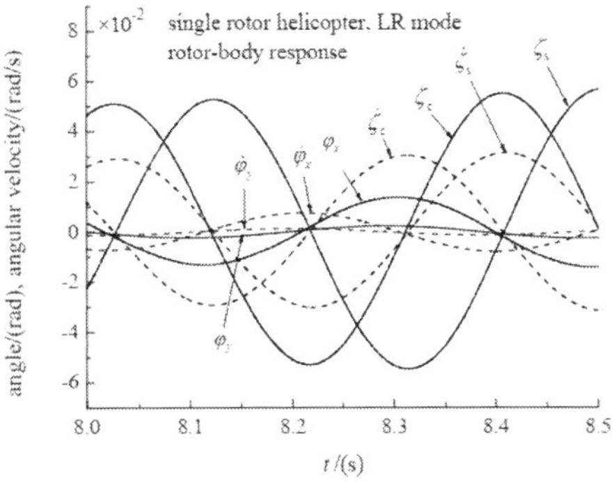

Figure 5. Response of various DOFs(Single rotor helicopter, LR mode, Ω=280r/min)

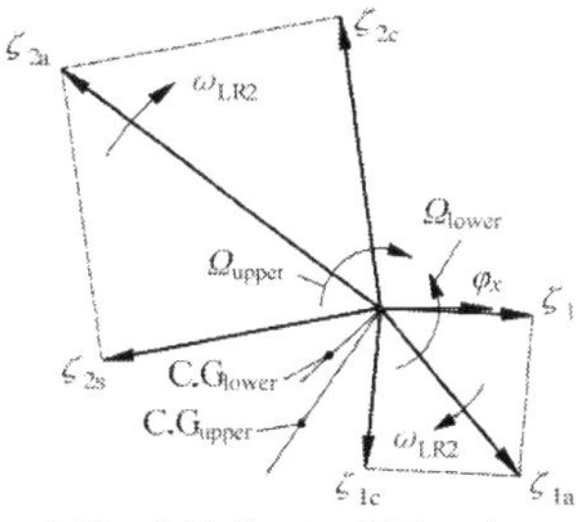

(a)Coaxial helicopter, LR2 mode

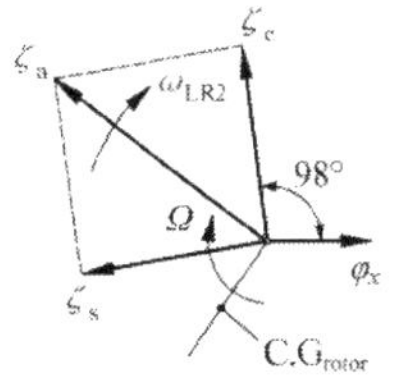

(b)Single rotor helicopter, LR mode

Figure6. Phase relation of various DOFs(Ω=280r/min)

Firstly, interaction between upper rotor(or single rotor) periodic lag and body roll motion of coaxial helicopter is analyzed.

Then, we can see from equation (11), for this upper rotor, there is only lag cosinoidal periodic component has inertia moment effect on body roll motion, Moment of inertia is: $M_{RB2}=-NS_b(h+h_x)\ddot{\zeta}_{2c}/2$,Thus, when $\ddot{\zeta}_{2c}$ and $\dot{\varphi}_x$ are in the same phase, the symbol of $M_{RB2}$and body roll damping moment is opposite. However, as can be seen from Figure 6, When $\ddot{\zeta}_{2c}$ (or $\ddot{\zeta}_c$) and ζ_{c2}(or ζ_c) are in the opposite phase, and be ahead of φ_x vector approaches to 90°. That is, when $\ddot{\zeta}_{2c}$ (or $\ddot{\zeta}_c$) and $\dot{\varphi}_x$ are close to the same phase, it provides a negative damping to body, namely, inputting energy to body.

Because of center of rotor gravity lags lag vector (Figure 6) 90°, and moment of body inertia and vector φ_x is in the opposite phase, body provides excitation to periodic lag motion of upper rotor, namely, inputting energy to lag of rotors. Lag sinusoidal component ζ_{2s} (or ζ_s) of upper rotor and body roll angle φ_x are close to opposite phase, so there is almost no energy transferred between φ_x and ζ_{2s} (or ζ_s), and they are mainly provided positive or negative "spring" constraint. Thus, for this LR2 mode (or single rotor helicopter LR mode), instability of helicopter ground resonance is mainly due to energy transferred between periodic lag motion of upper rotor and body roll rotation.

Next, interaction between lower rotor and body can also be analyzed combining with equation (11) according to phase relationships as shown in Figure 6(a). According to equation (11), for this lower rotor, there is only lag cosinoidal periodic motion has inertia moment effect on body roll motion, too, the size is $M_{\mathrm{RB1}} = NS_b h_x \ddot{\zeta}_{1c}/2$, when $\ddot{\zeta}_{1c}$ and $\dot{\varphi}_x$ are in the opposite phase, lower rotor provides negative damping to body; As rotary direction of lower rotor is opposite to that of upper rotor, and center of rotor gravity lags lag vector (shown in Figure 6) 90°, Therefore, moment body inertia has an incentive effect on periodic lag motion of upper rotor, that is, body transfers energy to rotors. In summary, for this LR2 mode, instability of helicopter ground resonance is mainly due to energy transferred between periodic lag motion of upper rotor and body roll rotation. For this stable region (such as Ω=350r/min), the response and phase relation of rotors and fuselage are shown in Figure 7 and Figure 8. It can be found that upper rotor and lower rotor center of gravity and body roll vector is almost in the opposite phase, periodic lag motion and body roll will not transfer energy to each other.

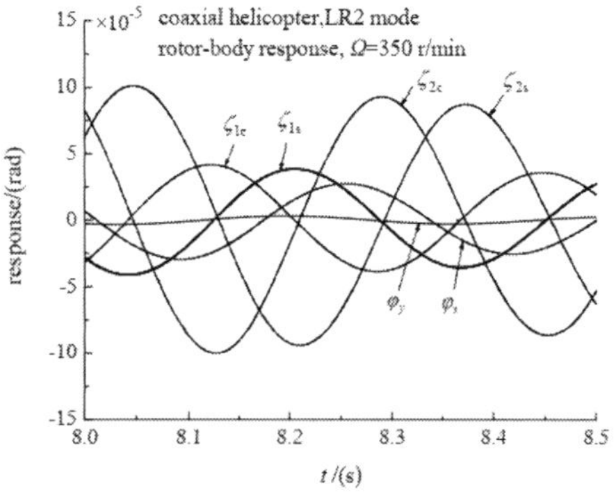

Figure 7. Response of various DOFs(LR2mode, Ω=350r/min)

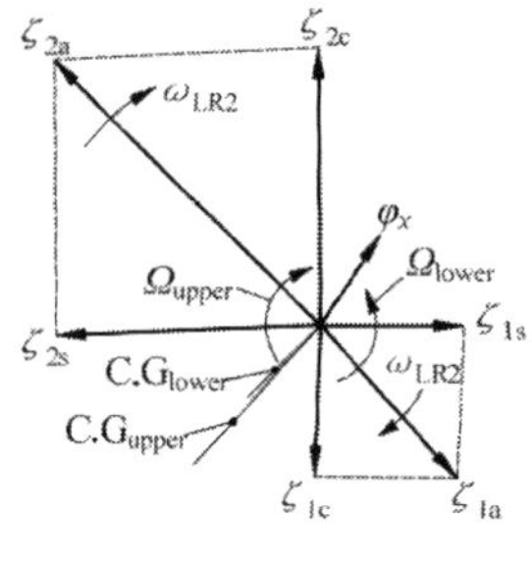

Figure 8. Phase relation of various DOFs(LR2mode, Ω=350r/min)

For the situation of body is isotropic(body pitch parameters are consistent with that of body roll), modal frequency and modal damping of system are quite different from those of the situation of body is anisotropy (Figure 9(a) and (b)), Providing that upper rotor turns positive, then, regressive lag LR2 mode with upper rotor characteristic and body mode of forward rotation B2 are coupled, and the modal damping is significantly lower than LR2 modal damping when body is anisotropy, it shows that dynamic instability of helicopter ground resonance is higher when body is isotropic, which is consistent with conclusion of analysis for single rotor helicopter ground resonance [1,2].

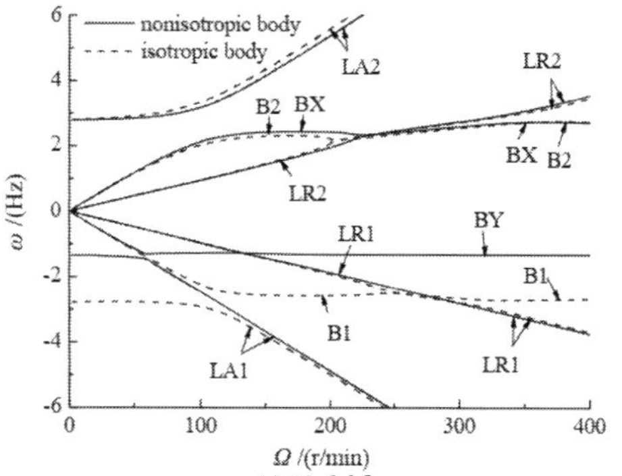

(a) Modal frequency

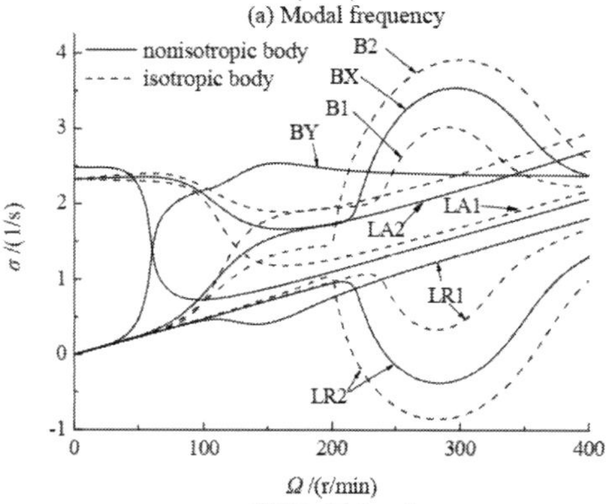

(b) Modal damping

Figure 9. Comparison of helicopter ground resonance modal characters between isotropic and nonisotropic body

Since pitch characteristic of body is in accordance with roll characteristic of body, frequency of body B1 mode is higher than frequency of pitch model (BY) when body is anisotropy, however, its rotary direction is the same as that of lower rotor, and it coupled with regressive lag LR2 mode with lower rotor characteristic, so that LR1 modal damping is significantly lower than damping when body is anisotropy, but is much higher than LR2 modal damping. Whether body is anisotropy or isotropic, there is both forward and reverse rotation of body mode of coaxial helicopter, however, there is only forward rotation when body of single rotor helicopter is isotropic. When body is isotropic, LR2 modal damping is significantly lower than damping value when body is anisotropic, which can be explained based on its response characteristic and phase relationship. On the other hand, when body is isotropic, for this unstable region(Ω=280r/min), periodic lag motion of rotor and response process of body are shown in Figure 10, phase relationship of various DOFs are shown in Figure 11 (the amplitude of periodic lag motion of lower rotor is very small and is not shown).

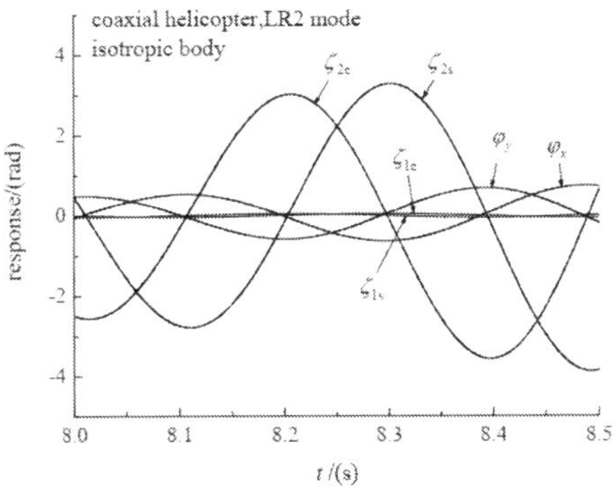

Figure 10. Response of various DOF(mode LR2, isotropy)

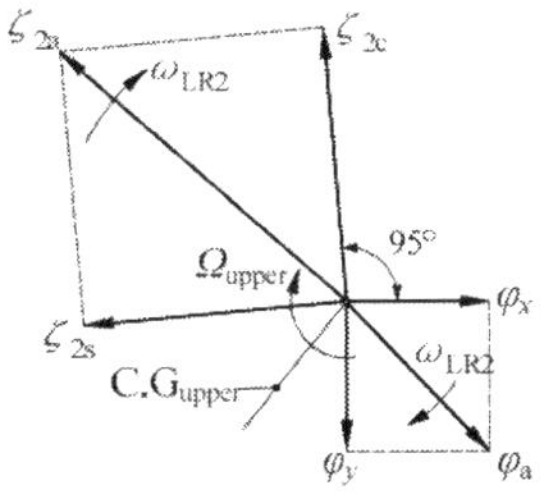

Figure 11. Phase relation of various DOF(mode LR2, isotropy)

As can be seen from Figure 10 and 11, when body is isotropic, for LR2 mode, periodic lag motion of upper rotor and body roll motion and pitch motion form a relationship of energy transferred, when body is anisotropy, energy transferred between rotor and body is increased, thus, increasing instability of helicopter ground resonance. While the amplitude of periodic lag motion of lower rotor is relatively small and almost be ignored, and this feature is obviously different from the situation when body is anisotropic.

As for the instability of coaxial helicopter ground resonance, it is also need to observe that the size of energy transferred between rotor and body and energy of structural damping consumes. Taking lag of lower rotor and upper rotor and body roll motion as the object of study as before, because one DOF provide negative damping to another DOF, in fact, it is doing positive power, so upper rotor and lower rotor power on body $W_{RB2}$and $W_{RB1}$are:

$$W_{RB2} = -\int_t M_{RB2}\cdot\dot{\varphi}_x \mathrm{d}t = \frac{1}{2}NS_b(h+h_x)\int_t \ddot{\zeta}_{2c}\dot{\varphi}_x \mathrm{d}t \quad (15)$$

$$W_{RB1} = -\int_t M_{RB1}\cdot\dot{\varphi}_x \mathrm{d}t = -\frac{1}{2}NS_b h_x\int_t \ddot{\zeta}_{1c}\dot{\varphi}_x \mathrm{d}t \quad (16)$$

And body powers on upper rotor and lower rotor $W_{BR2}$and $W_{BR1}$are:

$$W_{BR2} = -\int_t M_{BR2}\cdot\dot{\zeta}_{2c}\mathrm{d}t = S_b(h+h_x)\int_t \ddot{\varphi}_x\dot{\zeta}_{2c}\mathrm{d}t \quad (17)$$

$$W_{BR1} = -\int_t M_{BR1}\cdot\dot{\zeta}_{1c}\mathrm{d}t = -S_b h_x\int_t \ddot{\varphi}_x\dot{\zeta}_{1c}\mathrm{d}t \quad (18)$$

UsingΔW_1 to show the sum of power transferred between lower rotor and body, this is $\Delta W_1=\Delta W_{RB1}+\Delta W_{BR1}$, and using ΔW_2 to show the sum of power transferred between lower rotor and body, this is $\Delta W_2=\Delta W_{RB2}+\Delta W_{BR2}$. In unstable region of LR2 mode, energy transferred between periodic lag motion of rotors and body roll rotation was observed. In order to minimize the impact of other modes, starting to calculate energy transferred after 8 second, and the results show in Figure 12 and Figure 13.

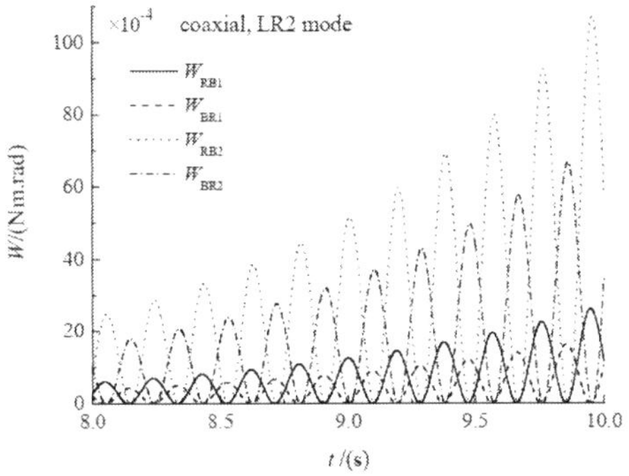

Figure12. Interaction history of rotors and body

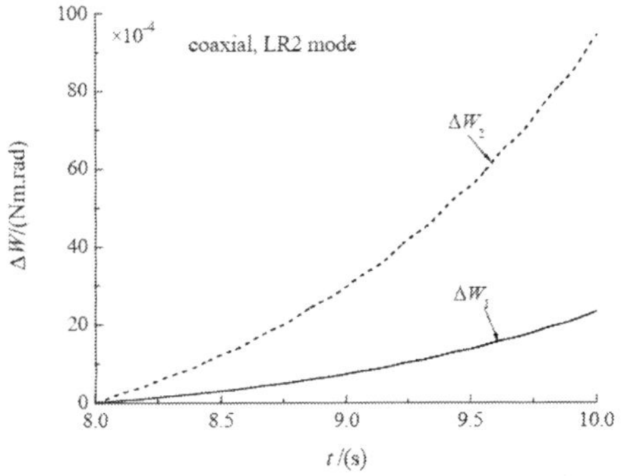

Figure 13. Increment of system power vs time

We can see from Figure 12, it is positive power transferred between periodic lag motion of upper rotor and lower rotor and body roll rotation, and it shows characteristic of alternating change with time, Its frequency is equal to that of LR2 mode, and body does not power on rotors when the maximum moment of rotors power on body, in addition, rotors do not power on body when the maximum moment of body powers on rotors, it shows accumulative power of coaxial helicopter rotor and body will increase with time in Figure 13, and the effect of upper rotor is significantly greater than that of lower rotor. Therefore, for LR2 mode, the main cause of coaxial helicopter ground resonance is interaction between periodic lag of upper rotor and body, while lower rotor enhances instability of coaxial helicopter ground resonance, this explains the intrinsic reason that dynamic instability of coaxial helicopter ground resonance is higher than that of single rotor helicopter. Of course, although the amplitude of body pitch motion is much smaller than that of body roll motion, considering interaction between body pitch motion and rotors, there maybe some change of value, but it will not affects the final results of analysis.

4Conclusion

(1) A time-frequency analysis method is proposed to reveal investigation of coaxial articulated helicopter ground resonance with structural damping. For mode with the least damping, interaction between periodic lag of upper rotor and lower rotor and body motion is analyzed by applying response characteristics of time-frequency, and internal reason of dynamic instability of coaxial helicopter ground resonance is explained.

(2) When body of coaxial helicopter is anisotropic, regressive lag LR2 mode with upper rotor characteristic is coupled with body roll mode, and it arouses body roll motion with the same direction of upper rotor rotation. While regressive lag LR1 mode with upper rotor characteristic and LR2 is in the opposite direction of rotation, there is not coupling between LR1 and body roll motion of reverse rotation, thus, helicopter ground resonance is caused by the coupling of regressive lag LR1 mode and body roll motion.

(3) In unstable rotary speed region of LR2 mode, coaxial helicopter ground resonance is mainly due to energy transferred between periodic lag motion of upper rotor and body roll rotation; In addition, energy transferred between periodic lag motion of lower rotor and body roll rotation improve dynamic instability of ground resonance.

(4) When body is isotroypic, there are two body modes B1 and B2 of coaxial helicopter that their frequency are similar and rotate reversely, this is significant difference from single rotor helicopter with only one body mode. B1 is the same as rotation of lower rotor and coulped with LR1 mode, and B2 is the same as rotation of upper rotor and coulped with LR2 mode, damping of LR2 mode is significantly lower than damping value when the fuselage is anisotropic.

(5) When body is isotropic, there is energy transferred between periodic lag motion of upper rotor and body roll rotation in unstable rotary speed region of LR2 mode, and instability of coaxial helicopter ground resonance is improved obviously comparing with body is anisotropy. While the effect of lower rotor and body motion is almost ignored.

References

1. Coleman M L, Feingold A M, Theory of self-excited mechanical oscillation of helicopter rotor with hinged blades, NACA Report TR-1351 (1958)
2. Zhang X G, Physical understanding of helicopter air and ground resonance, J. AHS, **31**,4,4-11(1986)
3. Ormiston, R A, Rotor-fuselage dynamics of helicopter air and ground resonance,J. AHS, **36**, 2,3-20(1991)
4. Russell E, Jennie S, Apache Helicopter Stabilization Using Neural Dynamic Programming, J. Guidance, Control&Dynamics, **25**, 1, 19-25(2002)
5. W Yuan, J Katupitiya, A Two-Stage Method for the Parametric Identification of Scale-Model Helicopter Dynamics, *Proceedings ofAIAA Atmospheric Flight Mechanics Conference,* 4733(2012)
6. Giuseppe Q, Vincenzo M, Pierangelo M. Lead-Lag Damper Robustness Analysis for Helicopter Ground Resonance, J. Guidance, Control&Dynamics, **36**, 4, 1150-1161(2013)
7. Tang D M, Dowell E H, Influence of nonlinear blade damping on helicopter ground resonance stability, J. Aircraft, **23**,2, 104-110(1986)
8. Smith E C, Govindswamy K, Beale M R, Aeroelastic response and stability of helicopters with elastomeric lag damper,J. AHS,**43**, 3, 257-266(1996)
9. Sahasrabudhe V, Gold P J. Reducing rotor-body coupling using active control, *Proceedings of AHS 60th Annual Forum*,1118-1834(2004)
10. Zhao Y S, Choi Y T, Wereley N M, Semi-active damping of ground resonance in helicopter using magnetorheological dampers, J. AHS, **49**, 4, 468-482(2004)
11. Bousman W G, An experimental investigation of the effects of aeroelastic couplings on aeromechanical stability of a hingeless rotor helicopter,J. AHS, 26, 1, 46-54(1981)
12. Liu L, Anand V R, Keith R H, Friedrich K S, Ground Resonance and Whirl Stability Analysis Using CAMRAD, *Proceedings of AHS 70th Annual Forum*,2586-2596(2014)
13. McNulty M I, Bousman W G, Integrated technology rotor methodology assessment workshop, NASA CP-10007(1988)
14. Wang, J.M., Duh,J, Fuh,J, Stability of the Sikorsky S-76 bearingless main rotor, *Proceedings of AHS 49th Annual Forum*, 983-1009(1993)
15. Krysinski T, Ferullo D, Overview of the EC155 dynamics validation program from design stage up to certification, *Proceedings of AHS 55th Annual Forum*, 1056-1063(1999)
16. Narramore J M, Mithat Yuce, Bell 429 main rotor aerodynamic and dynamic development, *Proceedings of AHS 66th Annual Forum*,436-452(2010)
17. Zhang X G, Liu Q, The application of complex coordinates and mutual excitation analysis to the investigation of helicopter ground resonance, Acta Aeronautica et Astronautica Sinica,**13**, 11, 586-593(1992)
18. Lu M, Fei B J, Ren Z Y, A numerical analysis for the ground resonance of a two-blade helicopter rotor on anisotropic flexible supports, Acta Aeronautica et Astronautica Sinica, **16**, 3, 348-354(1995)

Low-Cost and Aerodynamics-Aim Hypersonic Flight Experiment MF-1

Han-shan Xiao, Chao Ou [a], Hong-liang Ji, Zheng-chun He, Ning-yuan Liu and Xian-xu Yuan

China Aerodynamic Research and Development Center, China

Abstract. For increasing understanding of fundamental hypersonic phenomena, the flight test program, named MF-1, is to gather fundamental scientific and engineering data on the physics and technologies critical to future operational hypersonic flight with low-cost flight test platform, which is built on the retrofitted rockets. The MF-1 program is a hypersonic flight test program executed by China Aerodynamic Research and Development Center (CARDC). The MF-1 flight flew in December 2015. The flight focuses primarily on integration of instrumentation on the test vehicle, with application to boundary layer transition and shock interaction experiments. The MF-1 payload consists of a blunted 7° half angle cone, a cylinder and 33° flare configuration. The payload was boosted to Mach 5.32 utilizing a solid-rocket booster without control for the whole flight. The flight was fully successful, and measured transition under supersonic and hypersonic conditions. The heat flux data were given by the three-dimensional thermal identification method to discriminate transition zone. The preliminary analysis shows that the real-time flight data obtained by MF-1 are reliable and can be used to validate the transition predicting model and software. The results show that the existing model is able to predict the transition location of cone at a small angle-of-attack for supersonic or hypersonic flow. This paper describes the MF-1 mission and some general conclusions derived from the experiment.

1 Introduction

The MF-1 program is a hypersonic flight test program executed by China Aerodynamic Research and Development Center. MF-1 is the experimental flight with the primary experiment focused on boundary layer transition (BLT) on the low cost flight test vehicle. The corresponding secondary experiment is turbulent separated shock boundary layer interaction (SBLI). The research effort will consist of a series of focused tasks to resolve hypersonic phenomena through validation of computational analysis, comparison with performance predictions, enhancement of design data bases, and development of correlations with ground test[1].

Boundary layer transition has an important impact on hypersonic vehicle aerodynamics and aerothermodynamics. It is an inherently challenging problem due to its nonlinear nature and sensitivity to initial and boundary conditions. Ground test of boundary layer transition is generally unsatisfactory due to the high noise levels of ground facilities compared to flight. Computation is challenging since it requires resolution of instabilities of relatively high wavenumber and frequency over large spatial domains. Flight provides the best environment for measuring transition. Instrumentation for the flight environment however, is challenging. Careful measurement and computation of boundary conditions will provide added value in interpreting the data[2-3].

In recent years, a large number of flight tests have been carried out for hypersonic boundary transition and shock-border layer interference[4-5]. Such as the Hypersonic International Flight Research Experimentation (HIFiRE) program jointly implemented by the Air Force Research Laboratory (AFRL) of the United States and Australian Defence Science and Technology Group (DSTG), with a total of 3 flight tests dedicated to related issues (HIFiRE-1, 5, 9). As well as the Sharp Edge Flight Experiment program (SHEFEX) implemented by the German Aerospace Center (DLR), the SHEFEX-1 and 2 flight tests have been completed. Fig. 1 and Fig. 2 are typical research loads for HIFiRE and SHEFEX[6-13].

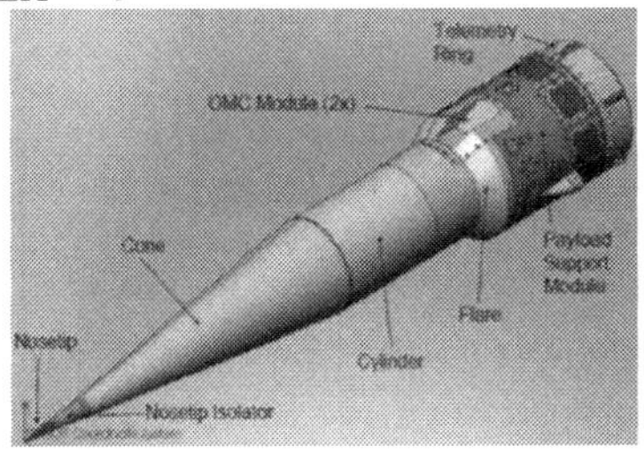

Figure 1. The payload of HIFiRE-1.

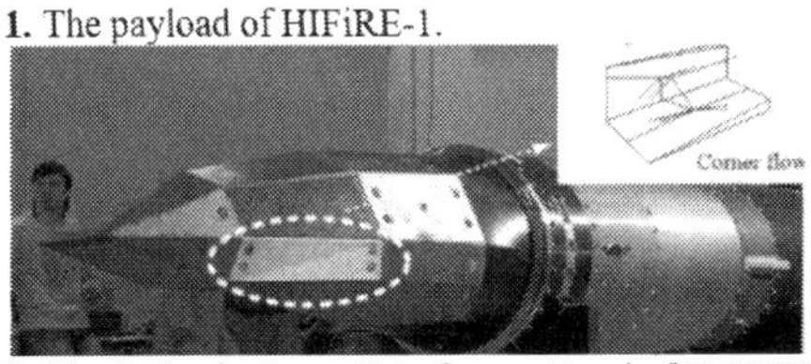

Figure 2. Shock-Border Layer Interference Load of SHEFEX.

[a] Corresponding author: hezhengchun@cardc.cn

The features which make the current experiment novel are the potential for hypersonic measurement, controlled roughness, or angle of attack measurements, and application of modern analysis. Improvements in modelling can now derive more consistent transition criteria from a flight experiment. In addition, historic experiments often did not record vehicle attitude or nose geometry, leading to increased uncertainty. Accurate knowledge of the vehicle attitude in the MF-1 flight is required to fully understand the BLT results.

MF-1 was launched on December 30 at 16:00:00 Beijing time in 2015. Telemetry data was received for the entire flight. This mission was entirely successful. This paper describes the test vehicle flight, initial post flight trajectory, vehicle attitude estimation, and some initial test results.

2 Flight configuration and integration

The overall payload dimensions and the different payload modules are shown in Figure 3 and Figure 4. The experiments are carried out on the forward sections of payload including a cone, a cylinder, and a flare. The cone half angle of seven degrees was chosen to match configurations used in preceding ground tests and analytical/numerical work (Fig. 5). The flare angle of 33° was chosen to induce turbulent boundary layer separation and reattachment on the flare face. To measure key aerothermal parameters during flight these two parts, which represent the main scientific part of the payload, were instrumented with a large number of sensors. Behind the flare experiment service modules with a length of 800 mm was used containing the necessary electronic boxes for data acquisition.

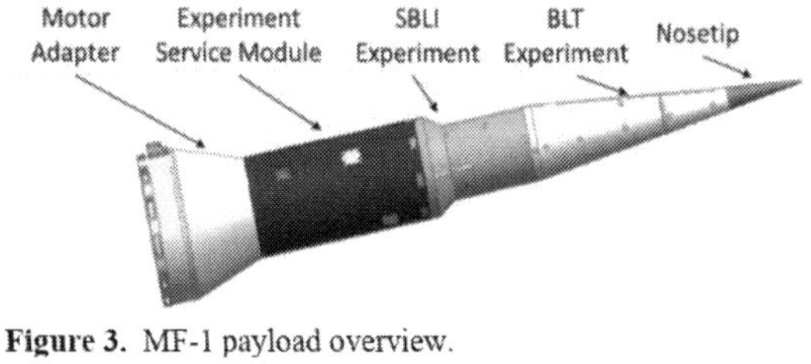

Figure 3. MF-1 payload overview.

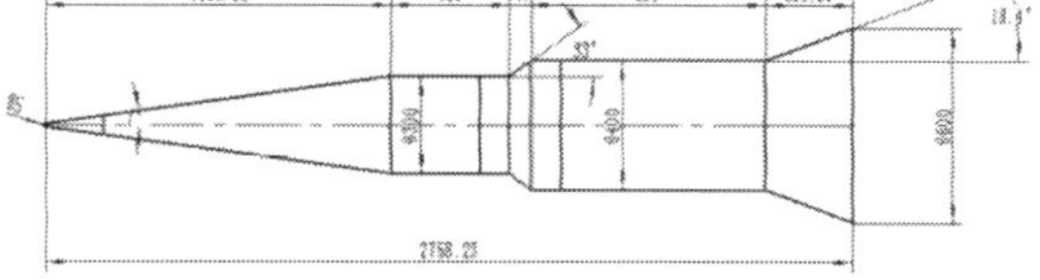

Figure 4. Payload geometry.

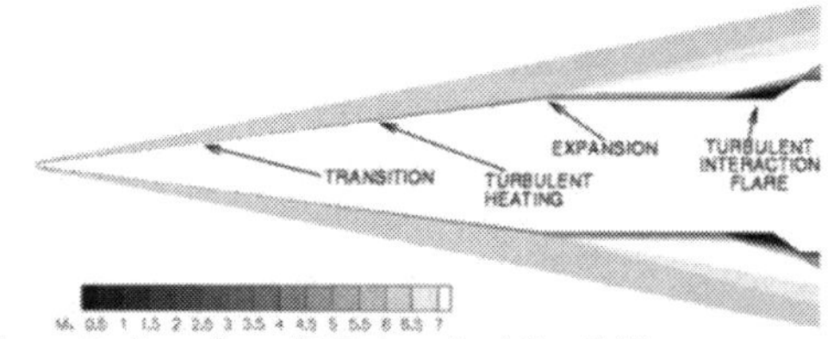

Figure 5. Overview of MF-1 payload flowfield.

The overall length of the payload including all service modules and motor adapter was 2758.23mm. This led to a total payload mass (including motor adapter and balance masses) of about 220 kg. The complete flight configuration with a length of 6.3 meter is shown in Fig. 6. The MF-1 vehicle was launched on an unguided rocket motor driven by solid propellant.

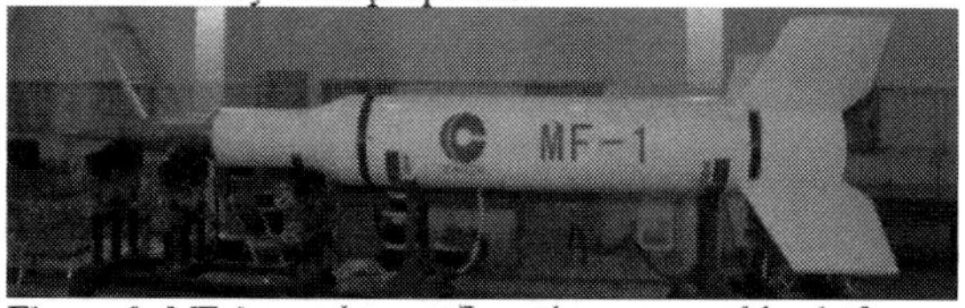

Figure 6. MF-1 complete configuration on assembly platform.

The main objective of the flight is to obtain the desired BLT and SBLI experimental data. For this to be achievable the payload must maintain its structural integrity until the end of the experiment. The payload had to be designed to withstand high aerodynamic heating loads resulting from hypersonic flight into a relatively dense atmosphere. The selected, parabolic, trajectory effectively limits the duration of in-atmosphere flight and consequent aero-thermal payload heating. As such it was possible to design the payload without having to add active or passive cooling systems and instead size the aero-shell walls to act as heat sinks. The nosetip stagnation region on the other hand reached temperatures approaching 1700K near the end of the trajectory. A high temperature alloy was used for the nosetip. The main body of the cone was made from stainless steel with a 12mm wall thickness. This thickness was chosen to provide sufficient heat sink mass for the duration of flight and still enable sufficient thermal response measurable with the chosen instrumentation. Located aft of the cone is a cylindrical section, required for isolation of the BLT and SBLI experiments. The cylinder section attaches to the cone section with blind fasteners, installed from the aft end of the cylinder. The cylinder was made from stainless steel and has a nominal wall thickness of 4.5mm. The flare section houses the SBLI experiment. The flare body was made from steel with 12mm to sustain higher temperatures on the flare face. The flare section attaches to the cylinder section using blind internal fasteners, similarly to the cone-cylinder joint. The payload on assembly platform is shown Fig. 7.

Figure 7. MF-1 payload on assembly platform.

The flight theory trajectory range was 224km, and the maximum altitude was 79.8km. The maximum flight Mach number was 5.53.

3 Instrumentation and data acquisition

3.1. Instrumentation

According to the theoretical prediction results of MF-1 boundary layer transition and shock boundary layer

https://doi.org/10.1051/matecconf/202031604006

interference, the temperature and pressure measurements on the surface of the payload were designed. The MF-1 flight experiment were equipped with overall 118 sensors, including 58 for temperature measurements, 60 for pressure measurements. The location of the instrumentation in cylindrical coordinates on the cone surface of the flight vehicle is shown in Figure 8. In this plot, $\phi=0°$ is a body-referenced azimuthal coordinate, with $\phi=0°$ coincident with the primary ray of thermocouple instrumentation. X axis is defined as the direction of the payload from the cone to the flare along the axis. The 58 temperature sensors were arranged at $\phi=0°$, 60°, 120° and 180°. The 60 pressure sensors were mainly concentrated near the cylinder-flare interface, to measure shock boundary layer interference. At x=800mm, 1400mm and 1580mm, 6 pressure sensors were arranged in the circumferential direction, the rest of sensors were equipped at $\phi = 90°$ and 270°, and the starting point of the pressure measuring points was located at x=400mm.

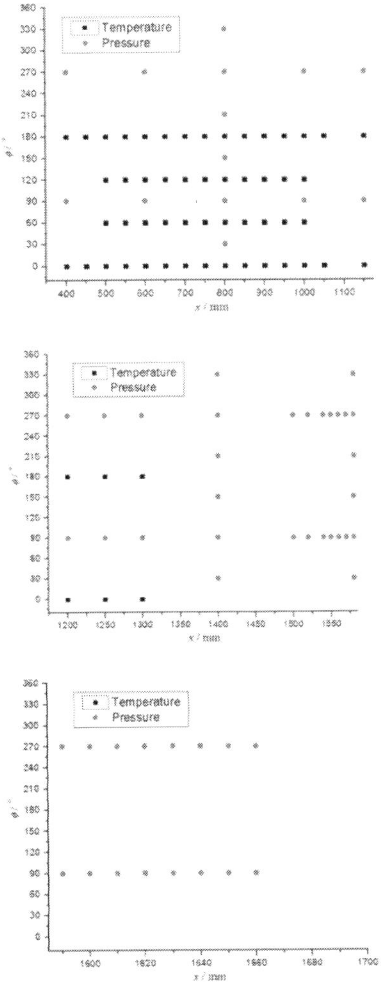

Figure 8. The location in cylindrical coordinates of the instrumentation on cone(up), cylinder(middle) and flare(down).

The optimization scheme of thin-skin model was used for temperature measurement structure. A thin-wall structure was designed for transition[14-15]. The measurement structure was shown in figure 9. The thermocouples were installed on backface of thin-wall. The outer surface temperature and heat flow data were obtained by data analysis and ground calibration. The thin-wall structure was welded with the payload structure. The solder joint has been polished to maintain aerodynamic shape after the welding was completed.

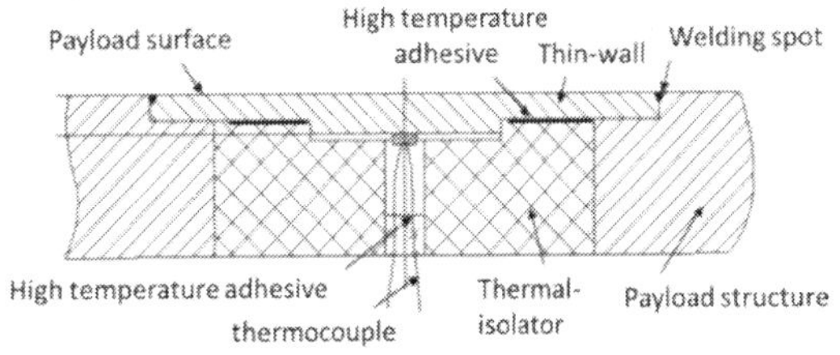

Figure 9. Thin-wall structure for temperature measurement.

The surface pressure measurement of payload used a pressure measurement system based on the electronic pressure scanner to measure 60 pressure data. A thin-wall structure was welded with payload at the measuring point. The solder joint has been polished after the welding was completed to maintain aerodynamic shape similarly to temperature measurement structure. Surface pressure was directed to the pressure scanner with stainless-steel tube and Polyvinylchlorid (PVC) tube. The structure of the pressure measurement is shown in Fig. 10

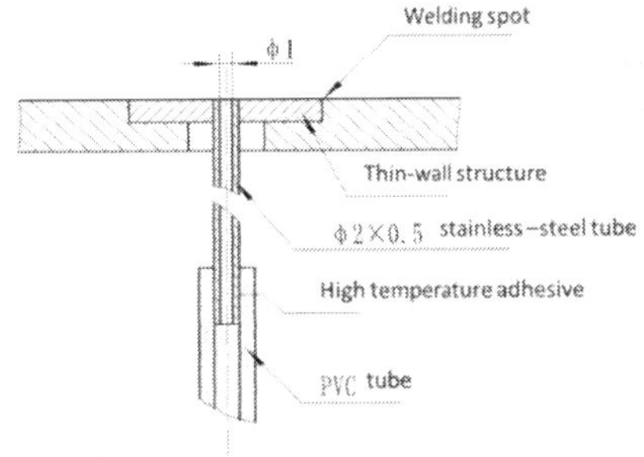

Figure 10. The surface pressure measurement structure.

3.2. Data acquisition

Data acquisition is carried out by means of telemetry emission recovery and data storage device recovery. The measurement data telemetry system collected sensor information such as temperature, pressure, Global Positioning System (GPS) and Inertial Measurement Unit (IMU) signal, and was received by ground equipment during the flight. Data storage device stored the measuring data during the flight and recovered the data after the flight test. The layout diagram of the telemetry equipment is shown Fig.11.

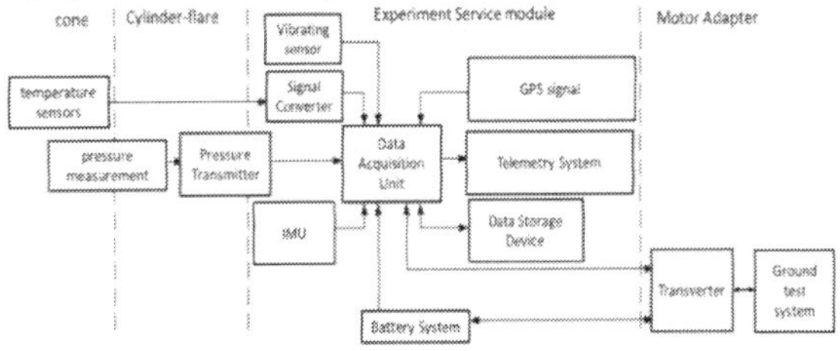

Figure 11. Data acquisition system of MF-1.

https://doi.org/10.1051/matecconf/202031604006

4. Flight data

4.1. Post flight trajectory

The MF-1 vehicle at the launcher is shown in Fig.12. The rocket stage accelerated the entire system with a total weight of 1739 kg to an altitude of 13.9 km as shown in Fig. 13. In the ascent phase the vehicle reached a maximum velocity of Ma 5.32 when the engine was shut off. The flight flew for 244.9 seconds. Data storage device was found after the flight. The debris of MF-1 is shown Figure 14.

Figure 12. MF-1 complete configuration at the launcher.

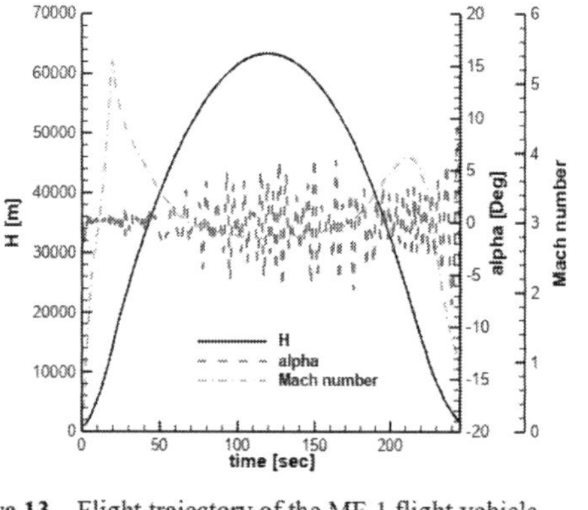

Figure 13. Flight trajectory of the MF-1 flight vehicle.

Figure 14. The debris of the MF-1 vehicle.

Fig. 15 shows a detailed view of Mach number (Ma) and Reynolds number (Re) during the experiment window within 10 to 30 seconds (ascent) and 210-230 seconds (descent). In the ascent, the flight Reynolds number decreases sharply due to the decrease in density as the altitude increases. However, due to the decrease of altitude and increase of density in the descending section, the flight Reynolds number gradually increases.

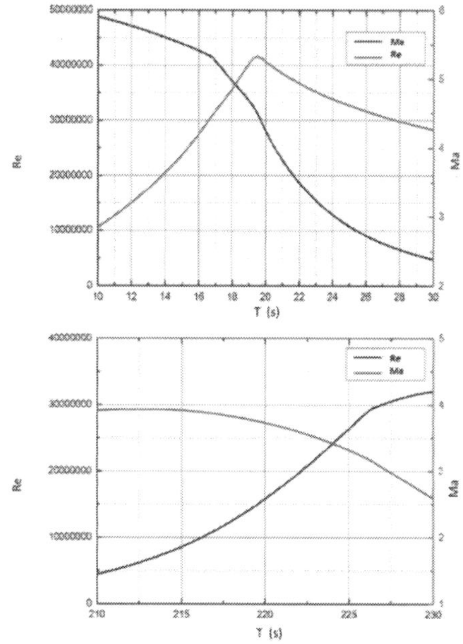

Figure 15. Flight Reynolds and Mach numbers during ascent(up) and decent(down).

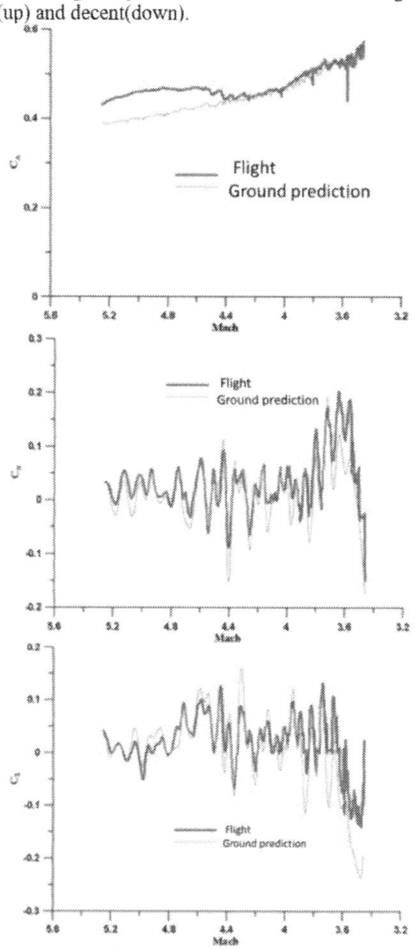

Figure 16. Comparison aerodynamic coefficient C_A(up), C_N(middle),and C_Z(down) between flight estimated and ground prediction.

The MF-1 flight data from data storage device was used to build an aerodynamic coefficient estimated. Compared this with the ground prediction data, and the results are shown in Fig. 16. When the Mach number was higher than 4.4, there would be a deviation in the axial aerodynamic coefficient (C_A) between flight results and the prediction of the ground, and the axial aerodynamic coefficient was higher in flight. The maximum deviation was more than 10%, but the trend was consistent. When the Mach number was less than 4.4 (22 km altitude above), The results of ground prediction were comparatively in agreement with flight. The normal aerodynamic coefficient (C_N) and lateral aerodynamic coefficient (C_Z) were consistent with the ground prediction results.

4.2. Heat Flux and Transition Result

The cone centerlines heat flux histories during the ascent and descent are shown in figures 17 and 18. As expected, the heat flux decreases as x increases and the boundary layer thickens. Jumps in heating rate due to transition are apparent at t = 16–22 and 215–223 s. The transition time remain basically unchanged along the cone surface at ϕ =0° between x =450 mm and x =1150 mm. But, there are significant changes in transition time along the cone surface at ϕ = 60°,120°,180°. At ϕ =0°centerline, an unchanged transition time can be explained by roughness-induced transition causing turbulence on the leading edge to advance forward rapidly when the critical Reynolds number is reached.

To reveal greater detail than in the heating histories, the heating rates along the centerlines located ϕ =0°,60°,120°,180 ° during the ascent and descent are shown in figures 19 and 20. The transition position changes significantly at different times along the cone surface at ϕ =60°,120°,180 °. An unaltered point of transition is also considered to be a forced transition. The reason may be due to the assembly step between nosetip and cone.

4.3. Measured Surface Pressure Result

Although measuring transition (via temperature) was the primary objective of MF-1, the vehicle was instrumented with numerous pressure sensors to measure shock interaction as well. The measured pressure of flare surface at ϕ =90 ° during ascend is shown in Fig. 21. However, the measured data may be inconsistent with the actual situation due to the improper measurement method. The measurement compared with Hifire-1 is shown in Fig. 26. The pressure oscillation phenomenon caused by shock boundary layer interaction did not appear in MF-1. The reason may be thought to be that the pressure measure tube was too long. The further details will be analysed and published.

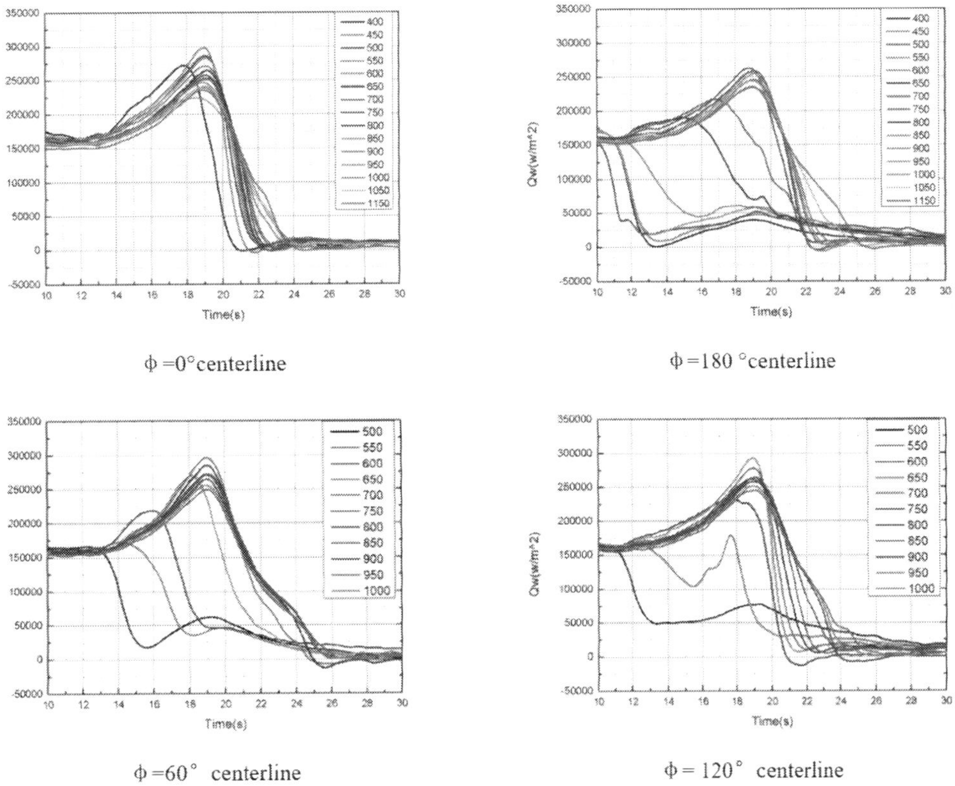

Figure 17. ϕ =0°,60° ,120° and 180° centerlines heat flux during ascent(t=10-30s).

https://doi.org/10.1051/matecconf/202031604006

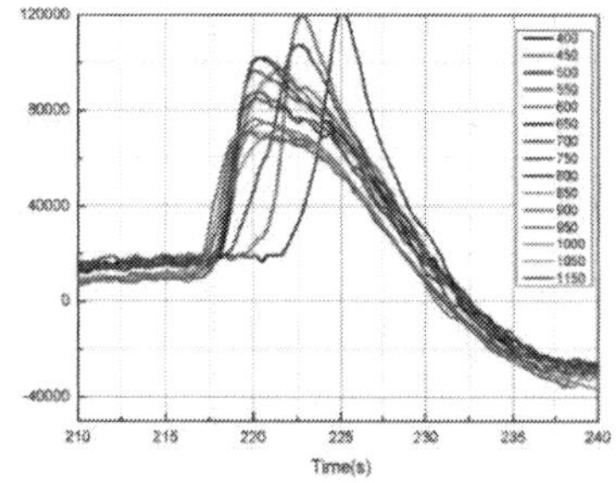

ϕ =0 ° centerline

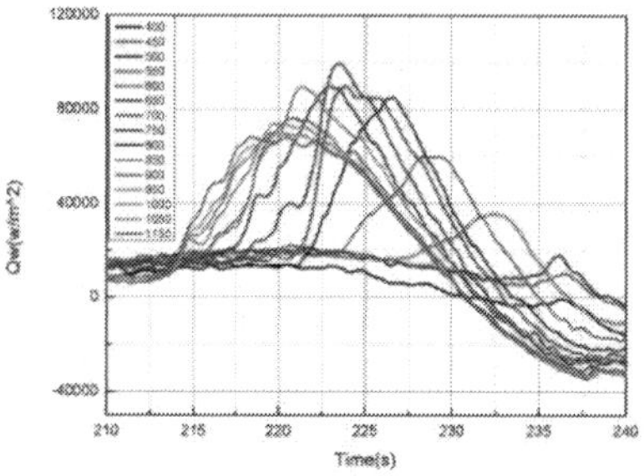

ϕ =180 °centerline

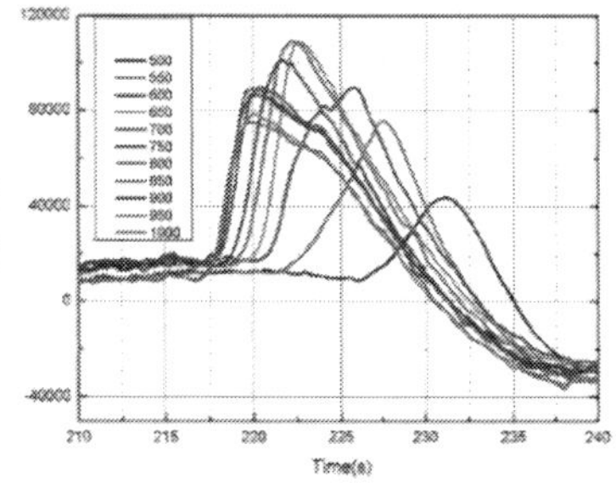

ϕ =60 ° centerline

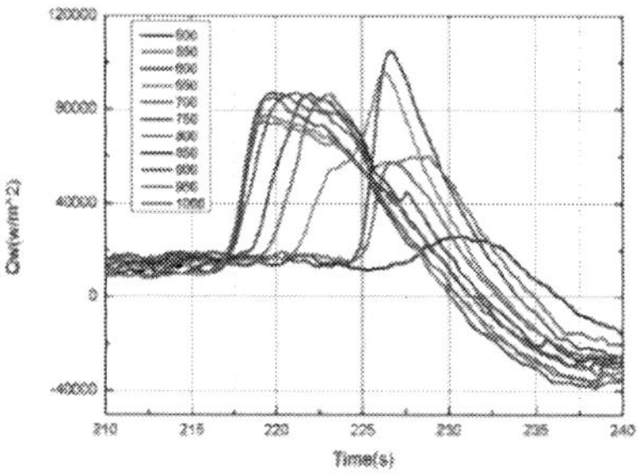

ϕ =120 ° centerline

Figure 18. ϕ =0°,60° ,120° and 180° centerlines heat flux during decent(t=210-230s).

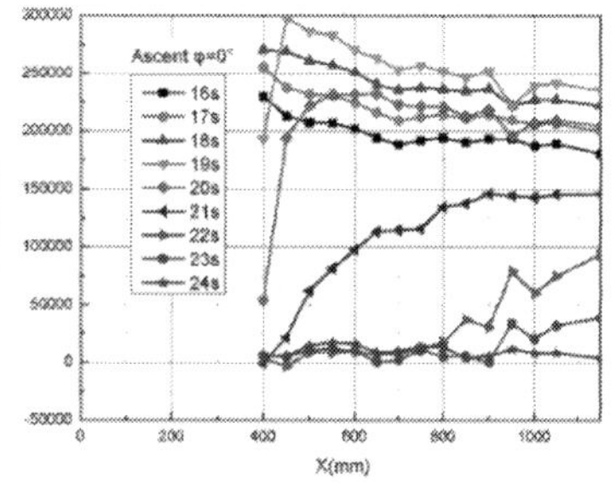

ϕ =0 ° centerline

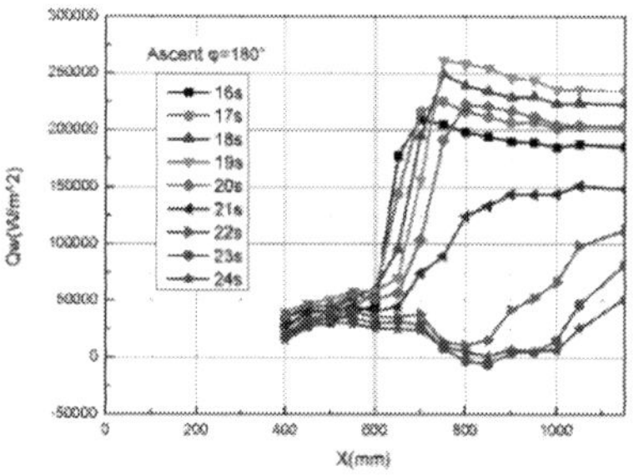

ϕ =180 ° centerline

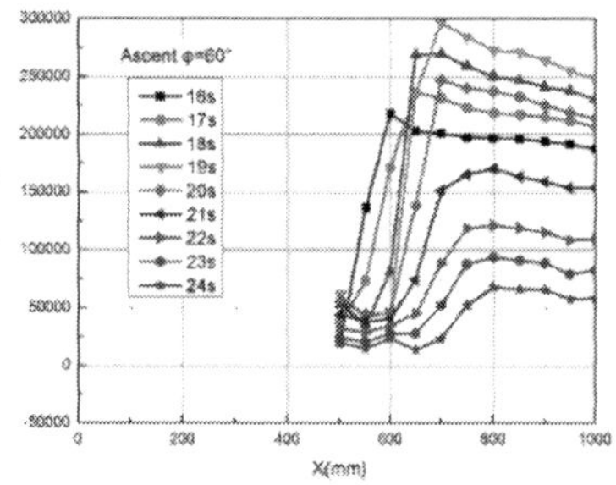

ϕ =60 ° centerline

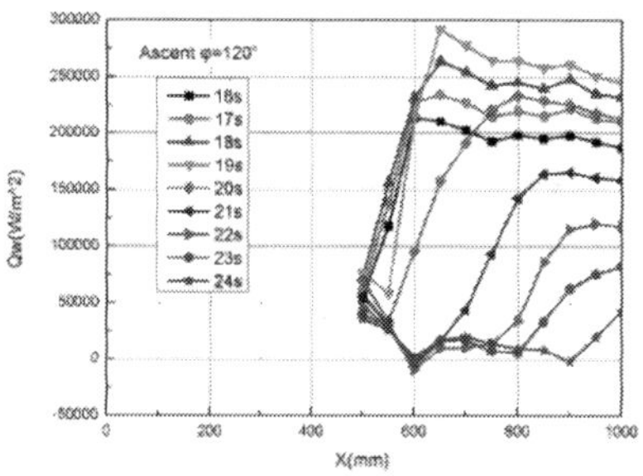

ϕ =120 ° centerline

Figure 19. ϕ =0°,60° ,120° and 180° centerlines transition during ascent(t=16-24s).

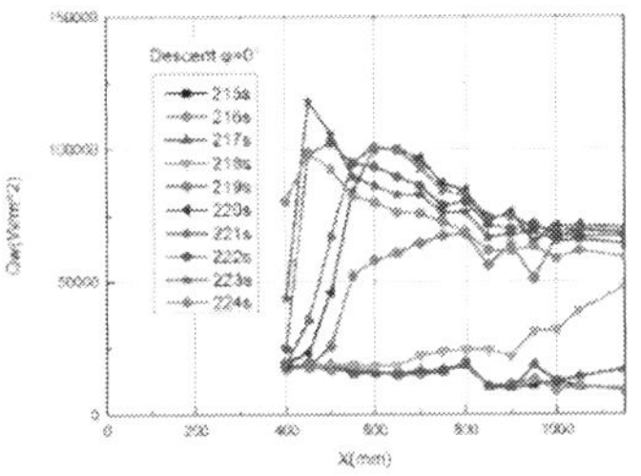

ϕ =0 °centerline

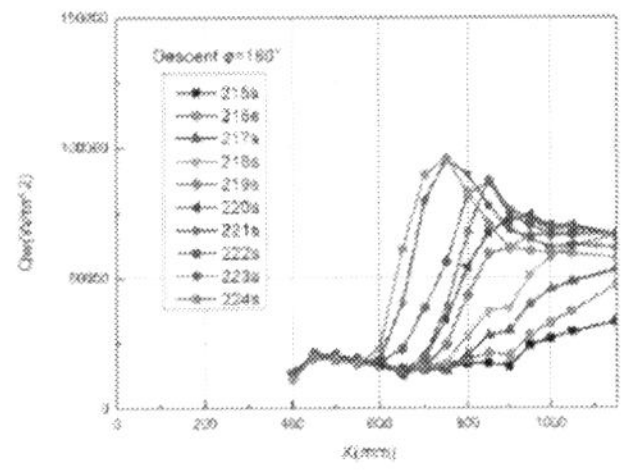

ϕ =180 °centerline

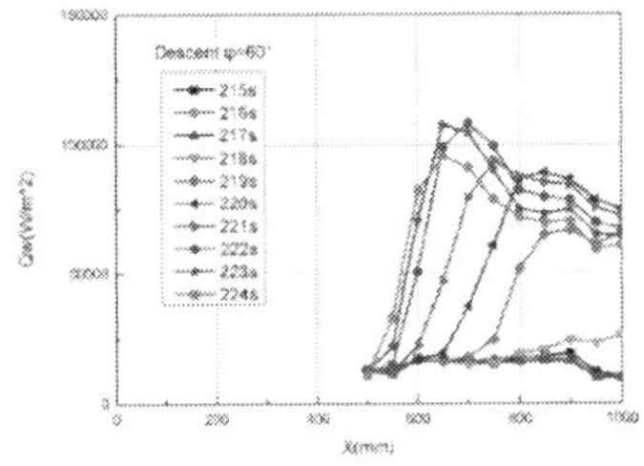

ϕ =60° centerline

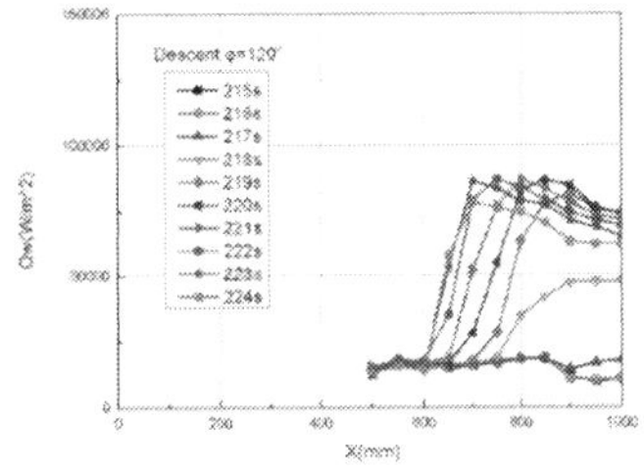

ϕ =120° centerline

Figure 20. ϕ =0°,60° ,120° and 180° centerlines transition during decent(t=215-223s).

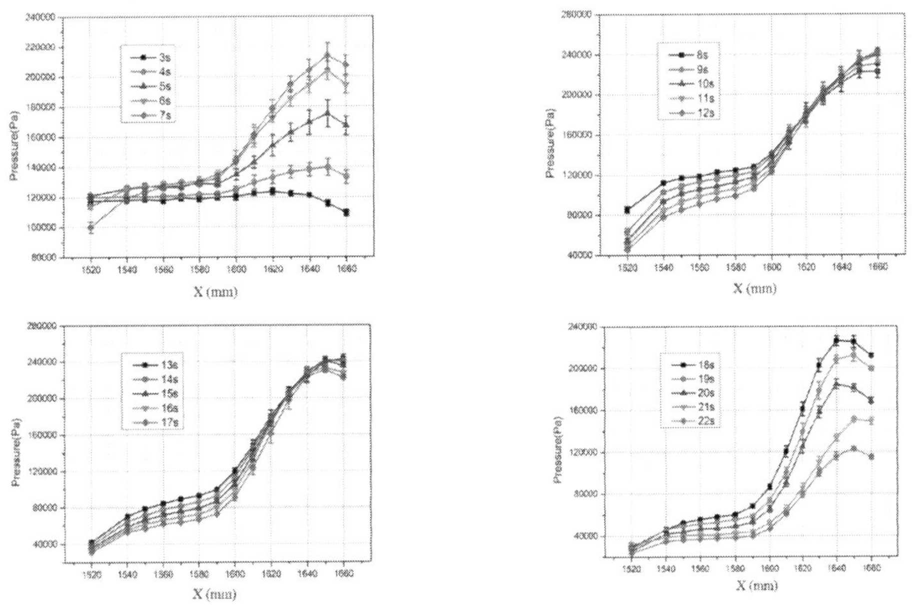

Figure 21. ϕ =90°centerline flare surface pressure during ascent.

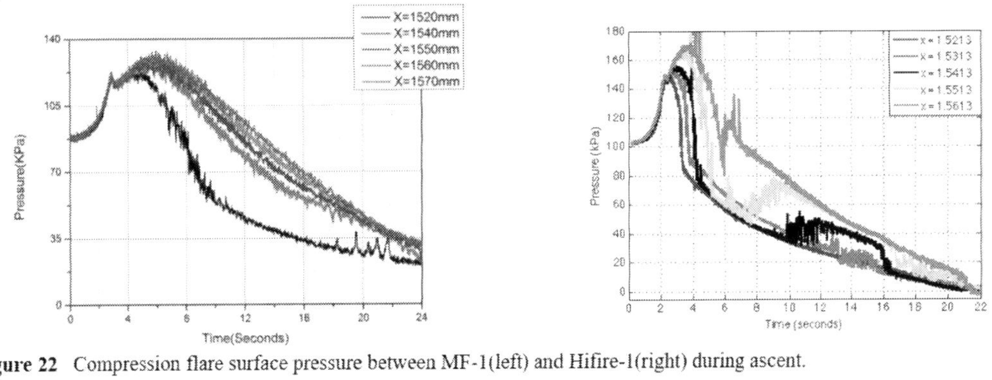

Figure 22 Compression flare surface pressure between MF-1(left) and Hifire-1(right) during ascent.

5 Conclusion

MF-1 was a highly successful flight for a "low-cost" flight experiment, 3+Gigabytes of data were obtained on boundary layer transition, shock boundary layer interactions as well as vehicle state data. Although MF-1was carried out as a low cost flight experiment without Control System and parachute system, the following important achievements were made:

-The launch, boost, tail and payload were carried out without any anomaly.

-The data acquisition systems worked perfect. The data storage device also worked well and all flight data were recovered. Most of the MF-1 instrumentation provided very useful data. The data can be used to verify the ground prediction method and calibration calculation model.

References

1. Ou C., JI H. L, Xiao H.S., et al. Key problems in structure and thermal protection for MF-1 model testing flight vehicle [J]. Acta Aerodynamica sinica, 2017,**35**(5):742-749.
2. Yang Q T, Zhou Y, Yuan X X, et al. Surface pressure and temperature measurement technology in MF-1 modelling flight test [J]. Acta Aerodynamica sinica, 2017,**35**(5):732-741.
3. He Z C, CHE J, XIAO H S, et al. Trajectory design and Monte Carlo flight simulation for MF-1 [J]. Acta Aerodynamica sinica, 2019,**37**(2):201-206.
4. Shehret Tilvaldyev, Erwin Martinez, Jorge Flores-Garay, and Alfredo Villanueva Montellano, "Distribution of Dynamic Pressure in Micro-Scale of Subsonic Airflow around Symmetric Objects at Zero Angle of Attack," International Journal of Mechanical Engineering and Robotics Research, Vol. **6**, No. 3, pp. 226-231, May 2017. DOI: 10.18178/ijmerr.6.3.226-231.
5. Yury I. Dimitrienko, Mikhail N. Koryakov, and Andrey A. Zakharov, "Computational Simulation of Conjugated Problem of External Aerodynamics and Internal Heat and Mass Transfer in High-Speed Aircraft Composite Construct," International Journal of Mechanical Engineering and Robotics Research, Vol. **6**, No. 1, pp. 58-64, January 2017. DOI: 10.18178/ijmerr.6.1.58-64
6. Kimmel, R. L. Aerothermal Design for the HIFiRE-1 flight Vehicle[C]. AIAA 2008-4034, June 2008.
7. Adamczak, D., Alesi, H., Frost, M. HIFiRE-1: Payload Design, Manufacture, Ground Test, and Lessons Learned[C]. AIAA paper 2009-7294, October 2009.
8. Roger L. Kimmel, David Adamczak. Hifire-1 Background and Lessons Learned[C]. AIAA paper 2012-1088, January 2012.
9. Schneider, S. P. Flight Data for Boundary-Layer Transition at Hypersonic and Supersonic Speeds[J]. Journal of Spacecraft and Rockets, vol. **36**, no. 1, January-February 1999.
10. Stetson, K. F. Nosetip Bluntntess Effects on Cone Frustum Boundary Layer Transition in Hypersonic Flow[C]. AIAA paper 83-1763, July 1983.
11. Kimmel, R. L., Adamczak, D., Gaitonde, D., Rougeux, A., Hayes, J. R., HIFiRE-1 Boundary Layer Transition Experiment Design[C]. AIAA paper 2007-0534, January 2007.
12. Johnson, H. B., Alba, C. R., Candler, G. V., MacLean, M., Wadhams, T, and Holden, M. Boundary Layer Stability Analysis of the Hypersonic International Flight Research Transition Experiments[C]. AIAA Journal of Spacecraft and Rockets, vol. **45**, no. 2, March-April 2008.
13. Stainback, C. P. Effect of Unit Reynolds Number, NoseBluntness, Angle of Attack, and Roughness on Transition on a 5°Half-Angle Cone at Mach 8[C]. NASA TN D-4961, January 1969.
14. LIU CHU-ping,YANG Qing-tao， et al. Heatflux measurement in aerothermodynamics and thermal protection tests [M]. Beijing:National Defense Industry Press,2013.
15. ZHANG Shou-yan,HUI Yu-xin， et al. Modeling Free flight [M]. Beijing: National Defense Industry Press,2002.

 https://doi.org/10.1051/matecconf/202031604007

A Fast Estimation Method for Satellite in Orbit Management

Zhen Cui[1,2,a], Ye Ji[1,2], Bin Chen[1]and Yang Yang[1]

[1]Beijing Institude of Control Engineering, Beijing,100094, China
[2]Science and Technology on Space Intelligent Control Laboratory, Beijing,100094, China

Abstract. The management of satellites in orbit requires accurate and rapid processing to minimize the impact of failures. Especially for fault handling involving energy security, the real-time nature of the process plays a decisive role. This paper proposes a fast estimation method based on satellite telemetry data. Through the analysis of the shape of the telemetry curve, the geometric method, the correlation analysis of the telemetry data on the satellite, or the physical principles, the required parameters are quickly obtained to facilitate rapid emergency processing. This method is applied to satellite in-orbit management, which can greatly improve the real-time performance of fault processing, and has good engineering practical value.

1 Introduction

There are at least two differences between satellite in-orbit management and satellite manufacturing, that is, in-orbit operations have higher and more stringent requirements for correctness and real-time performance. Once the satellite's in-orbit operation is wrong or the best processing window is missed, the impact is often greater and the consequences are more serious. For example, in the ground test or simulation of a satellite, if an operation error causes the satellite to switch from normal mode to sun capture mode, the device will usually be restarted to continue testing, but if similar problems occur in orbit operation, satellite service will be interrupted. The consequences and impacts are serious. For another example, in ground testing or simulation, if the attitude diverges or the rotation angle of the solar array is wrong, it usually does not cause energy security problems. However, for satellites in orbit, especially those launched in the early days, satellites with weak autonomous functions, if errors cannot be quickly processed, or the best processing window is missed, it may cause serious consequences affecting satellite energy security.

In addition, the following methods are usually adopted for emergency treatment of in-orbit faults. First analyze the phenomenon, then program the simulation to obtain the required parameters, and then send them to the satellite for troubleshooting. This method has two problems in engineering practice. One is insufficient real-time performance, which makes it difficult to perform programming and simulation calculations at the fault processing site, and the other is that calculation errors sometimes occur. This leads to the so-called "accurate but incorrect" calculation result. Although the calculation result is accurate to a few digits after the decimal point, it is an invalid result, and the consequences of this situation are often very serious.

Some scholars have researched how to avoid the problem of "accurate but incorrect" during simulation and calculation. For example, Lu Jiaxi proposed a method of estimation in research [1]. By establishing a simplified model, a reasonable estimate of the calculation results can be obtained, which can effectively avoid "precise errors" in analytical calculations. Yang Weilian compares "simulation" and "approximate" [2], and illustrates the significance and importance of "approximate" in engineering through some examples. Yang Weilian mentioned: "Simple models are easier to understand and easier to check than complex models, so it is easier to ensure their correctness. Conversely, the more complex the model, the more likely it is to make mistakes in the application process."

This paper proposes a fast estimation method based on satellite telemetry parameters, which is an engineering approximate calculation method. This method can improve the accuracy and real-time performance of in-orbit satellite management, and has certain engineering value. This method can effectively avoid "precise errors" in analytical calculations by establishing a simplified model and quickly estimating the reasonable range of the parameters. This is exactly what Keynes calls "Better be roughly right than precisely wrong".

2 Fast estimation method

2.1 Usage of estimation methods

[a] Corresponding author: 5277981@qq.com

In satellite in-orbit management, the application of the fast estimation method is roughly derived from the following requirements.

1) Accurate calculations often require complex formulas or programming, and they are often not equipped for satellite emergency processing.

The occurrence of satellite failures is often unpredictable and may occur at any time. This requires emergency treatment in the field, and it is difficult to have conditions such as programming simulation at this time. The estimation method can be quickly processed according to the relevant telemetry parameters on the satellite.

2) Satellite emergency processing usually requires rapid and effective processing methods, and does not necessarily require accurate results.

For example, when the satellite attitude deviates from the target attitude in orbit, and the solar array deviates from the target rotation angle, which affects satellite service or energy security, the attitude or solar array rotation angle needs to be adjusted quickly to ensure energy. At this time, the estimation method based on satellite telemetry parameters is often faster.

3) The estimation method often helps to quickly determine the cause of the fault.

When a satellite fails, comprehensive analysis is required to determine the cause of the failure. For example, when the attitude angle fluctuates, it is necessary to quickly analyze the attitude angle, the output of each sensor, the rotation angle of the solar array, the rotation speed of the reaction wheel, angular momentum, and the orientation of the sun and the moon. The estimation method helps to quickly analyze the parameters and quickly determine the cause of the failure.

4) The estimation method and the programming simulation method are compared with each other, which is helpful to improve the accuracy of in-orbit processing.

The estimation method is simple and fast, and the physical or geometric meaning is clear. The rationality of the simulation results can be judged to prevent the simulation calculation from appearing "accurate but incorrect" results. The simulation calculation can accurately estimate the error of the estimation method and prevent large errors in the estimation method.

2.2 Fast estimation based on curve topography

In-orbit management and troubleshooting generally need to be clear about the question of "where is the sun", that is, the elevation and azimuth information of the sun vector. If the programming simulation method is adopted, the representation of the sun direction vector in the orbit coordinate system needs to be calculated.

$$\mathbf{S}_o = \mathbf{A}_{ou}\mathbf{A}_{oi}\mathbf{S}_i \tag{1}$$

Where $\mathbf{S}_i$ is the representation of the sun vector in the inertial coordinate system. It is defined as:

$$\mathbf{S}_i = \begin{bmatrix} S_{ix} & S_{iy} & S_{iz} \end{bmatrix}^T \tag{2}$$

$\mathbf{S}_i$ needs to be calculated according to the ephemeris of the sun.

$\mathbf{A}_{ou}$ and $\mathbf{A}_{oi}$ are 3×3 transformation matrices related to the satellite orbit, $\mathbf{A}_{ou}$ is defined as:

$$\mathbf{A}_{ou} = \begin{bmatrix} \cos u & 0 & \sin u \\ 0 & 1 & 0 \\ -\sin u & 0 & \cos u \end{bmatrix} \tag{3}$$

$\mathbf{A}_{oi}$ is defined as:

$$\mathbf{A}_{oi} = \begin{bmatrix} \cos i \cos\Omega & -\sin i & \cos i \sin\Omega \\ \sin i \cos\Omega & \cos i & \sin i \sin\Omega \\ -\sin\Omega & 0 & \cos\Omega \end{bmatrix} \tag{4}$$

Where $u=\omega+f$, ω is the argument of periapsis, f is the true anomaly, i is the orbit inclination and Ω is the right ascension of ascending node[3].

$\boldsymbol{S_o}$ is the representation of the sun direction vector in the orbit coordinate system. The $\boldsymbol{S_o}$ can be calculated before the subsequent calculation of various angles. This process is relatively complicated, and it is difficult to judge the rationality of the results. Moreover, it is difficult to have the conditions to perform such calculations at the site of the fault treatment.

The following describes a method for quickly determining the solar position based on the shape of the satellite telemetry curve. According to this method, the solar position of the satellite at each time in a future orbit cycle can be quickly estimated and predicted.

For example, for IGSO navigation satellites [4-6], the continuous maneuver control of the yaw angle keeps the sun vector in the $X_bO_bZ_b$ plane of the satellite body coordinate system, and controls the rotation angle of the solar array so that the normal of the solar array always points in the direction of the solar vector [7-11]. The position of the sun includes the sun elevation angle θ_s and the sun azimuth angle α . The sun azimuth angle α is the angle between the projection of the sun vector on the orbital $X_oO_oZ_o$ plane and the $+Z_o$ axis, and it is positive from the $+Z_o$ axis $+X_o$ axis rotation. The elevation angle α of the sun is the angle between the sun vector and the orbital plane, with the sun vector pointing to the orbital system $+Y_o$ plane as positive.

The yaw angle curve of an orbital period contains the information of the solar elevation angle θ_s, as shown in Figure 1.

$$\theta_S = \min(\psi(t)) \tag{5}$$

Where $\psi(t)$ is the yaw angle of the satellite.

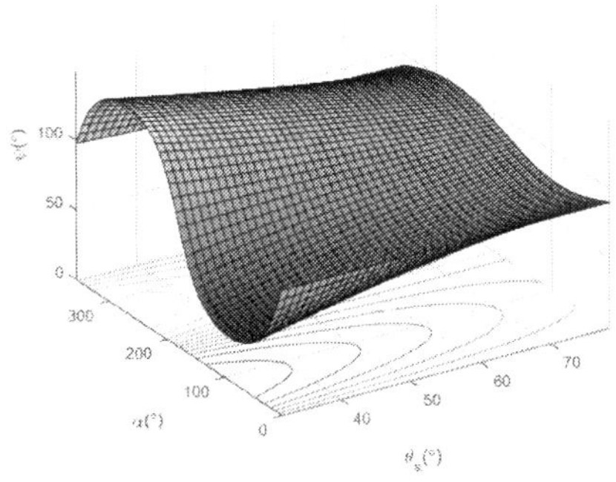

Figure 1. Variation of yaw angle with θ_S and α

The solar array angle curve of one orbital period contains the information of the solar azimuth angle.

$$\cos\alpha\cos\theta_S = \cos\left(\beta_{sada} - \pi\right) \quad (6)$$

β_{sada} is the angle of the solar array.

For example, for a GEO navigation satellite, since the control targets of the three-axis attitude are all 0°, the solar position can be estimated based on the output of each solar sensor or the rotation angle of the solar array.

The output angle curve of the solar sensors DSS13 and DSS43 in one orbit period contains the information of the solar elevation angle θ_S.

$$|\theta_S| = \min\left(\alpha_{DSS}(t)\right) \quad (7)$$

Among them, $\alpha_{DSS}(t)$ is the output angle information of digital solar sensors DSS13 and DSS43.

The solar array angle curve of one orbital period contains the information of the solar azimuth angle α.

$$\alpha = \beta_{sada} - \pi \quad (8)$$

2.3 Fast estimation based on geometric relations

The following uses an example in engineering practice to illustrate. In the in-orbit management of the MEO navigation satellite, the satellite is in a yaw dynamic offset mode. The continuous maneuver control of the yaw angle keeps the sun vector in the $X_oO_oZ_o$ plane of the satellite body coordinate system, and controls the rotation angle of the solar array so that its normal line always faces the direction of the sun vector. Assume that due to the need for phase adjustment, or the need for emergency treatment of the fault, the satellite needs to enter the phase adjustment mode and make the yaw angle 0°. At this time, how to adjust the rotation angle of the solar array so that it faces the sun as much as possible?

If the programming simulation method is adopted, the expression of the solar vector in the orbit coordinate system still needs to be calculated according to formula (1). In fact, the calculation of the space angle has nothing to do with the choice of the coordinate system. If the method of coordinate system and transformation matrix is not adopted, the telemetry parameters of the satellite can be obtained, and the simple and fast estimation can be performed through the geometric relationship.

It can be known from formula (5) that the solar elevation angle θ_S can be obtained from the yaw curve of the satellite in an orbital period, and the solar array rotation angle β_{sada} is a direct telemetry parameter of the satellite. Then through simple geometric calculations, the optimal rotation angle of the solar array can be obtained.

$$\cos\beta_0\cos\theta_S = \cos\left(\beta_{sada} - \pi\right) \quad (9)$$

Then the satellite yaw angle needs to be adjusted as follows.

$$\Delta\beta_{sada} = \beta_{sada} - \left(\pi + \beta_0\right) \quad (10)$$

Figure 2 shows the surface of $\Delta\beta_{sada}$ as it changes with θ_S and β_{sada}. $\Delta\beta_{sada}$ is monotonically decreasing or rising between θ_S and 0. When θ_S is large, the shape of the surface is equivalent "flat". In engineering practice, for example, in the case of emergency handling of a fault, or when it is related to energy security, if the simulation calculation conditions are not available at the fault processing site, $\Delta\beta_{sada}$ can estimate according to a linear method to adjust the rotation angle of the solar array to ensure satellite energy.

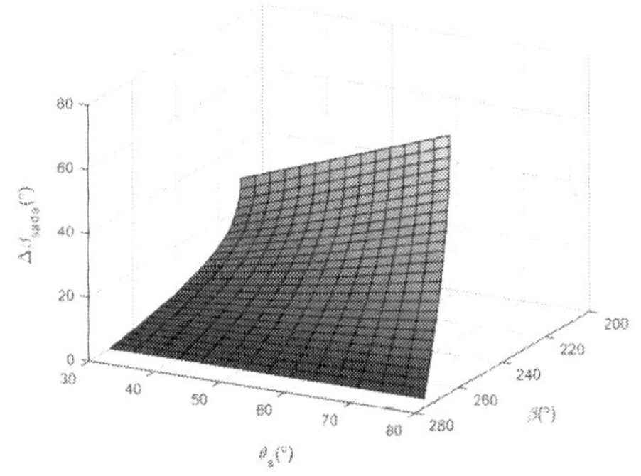

Figure 2. $\Delta\beta_{sada}$ changes with θ_S and β_{sada}

2.4 Fast estimation based on correlation of telemetry data

In the process of satellite in-orbit management and in-orbit fault processing, since we can directly obtain a large amount of telemetry data at the processing site, we should make full use of the correlation between these telemetry parameters for rapid estimation.

For example, an IGSO navigation satellite has entered solar capture mode due to incorrect orbit data. Because the fault occurs at night and there may be satellite eclipse, it is necessary to immediately inject the parameters related to satellite eclipse. However, since the failure occurred at night, it was difficult to carry out the simulation calculation immediately. At this time, a quick estimation can be performed based on the yaw angle of the previous orbit period and the rotation angle parameters of the solar array.

According to formula (5), determine whether the satellite is in the period of eclipse or not based on the curve shape of the yaw angle of the previous orbital period.

$$\theta_S = \min\left(\psi\left(t\right)\right) > R_e + R_s \qquad (11)$$

If formula (11) is satisfied, it is not in the period of eclipse, otherwise it is in the period of eclipse. Where R_e is the half cone angle of the earth and R_S is the half cone angle of the sun. If the satellite is in the eclipse period, the parameters related to the eclipse can be obtained by querying the solar array rotation telemetry data in the normal mode of the previous orbital period.

2.5 Fast estimation based on physical principles

Taking the failure of the momentum wheel on the satellite as an example, according to the principle of conservation of angular momentum, the shape of the speed curve of each momentum wheel after the failure and how to recover can be estimated. Neither simulation nor calculation is required, nor is it necessary to switch to earth pointing mode or position maintaining mode for processing. For example, a GEO navigation satellite is equipped with four pyramid-shaped momentum wheels to form a wheel control system [12]. If the entire satellite has zero momentum and a certain momentum wheel is powered off, since jet unloading is not allowed, the satellite keeps angular momentum conserved, so the speed of the other three momentum wheels will also decrease simultaneously. If the momentum wheel is restarted, and then the restarted momentum wheel is restored to the target speed, other momentum wheels will naturally be restored to the target speed simultaneously.

2.6 Comprehensive of fast estimation method

In engineering practice, especially for the emergency treatment of in-orbit faults, it is necessary to comprehensively apply the above-mentioned multiple methods for rapid emergency treatment. The comprehensive application of the rapid estimation method requires attention to the following:

1) When emergency handling of satellite in-orbit faults, the parameters are estimated based on the morphological characteristics of the curve, and the morphology of the curve in the recent orbital cycles before the failure should be identified. It is necessary to identify which is the correct data and which is the data at the time of failure. Try to avoid using only one orbit period of telemetry data for estimation, and prevent the use of wrong telemetry data.

2) In the case of non-emergency in-orbit processing, when the conditions for simulation calculation are available, simulation calculation based on ephemeris and rapid estimation based on telemetry parameters should be performed at the same time. Two different calculation methods are mutually checked and verified to avoid calculation errors. Fast estimation can avoid "accurate errors" in simulation calculations, and simulation calculations can prevent large estimation errors from occurring in rapid calculations.

3) When a fast estimation method based on the correlation of telemetry parameters is used, there are often multiple estimation methods for the same parameter, and multiple methods should be verified with each other to prevent estimation errors. For example, satellite attitude information, measurement information from multiple sensors, solar array rotation angles and other satellite telemetry data are all related to solar position information. Comprehensive verification should be performed through multiple methods and multiple associated telemetry data.

4) If the satellite attitude diverges, fast attitude recovery is needed, or the solar array angle needs to be adjusted quickly if the satellite energy security is endangered. When the simulation calculation conditions are not available, a fast estimation method based on telemetry parameters should be selected for processing. If the failure is not handled in a timely manner, it will lead to serious risks affecting satellite services or safety.

3 Conclusion

Aiming at the long-term in-orbit management of satellites and emergency handling of in-orbit faults, this paper proposes a fast estimation method based on on-board telemetry parameters and their correlation analysis, and illustrates the application value of this method through specific examples. Compared with the traditional programming or simulation method, this method can be quickly processed at the fault site, the calculation is simple, the geometric significance is obvious, and it can be applied to the practice of satellite in-orbit management.

References

1. Lu Jiaxi. On the "rough estimation" method in scientific research. Academician Thinking Volume III, [M],Hefei: Anhui Education Press, 2001: 130-136.
2. Yang Weilian. Simulation and approximation—discussion on engineering methodology. Proceedings of the 2004 Academic Symposium of the Aircraft General Professional Committee of China Aerospace Society, 2005, 133-136.

3. Zhang Renwei. Satellite Orbit Attitude Dynamics and Control [M]. Beijing: Beijing University of Aeronautics and Astronautics Press, 1998:6-8.
4. Mao Yue, Song Xiaoyong, Wang Wei, et al. Beidou IGSO and MEO navigation satellites' yaw-steering and orbit-normal attitude control modes and solar radiation pressure difference analysis[J]. Science of Surveying and Mapping, 2015, **40** (8) : 129-134.
5. Mao Yue, Song Xiaoyong, Wang Wei, et al. IGSO satellite orbit determining strategy analysis with the yaw-steering and orbit-normal attitude control mode switching[J]. Geomatics and Information Science of Wuhan University, 2014, **39** (11) : 1352-1356.
6. Li Xiao Jie, Liu Xiao Ping, Xin Jie, et al. Attitude Control Modes and Characteristic Analysis for BeiDou Satellites[J]. Journal of Geomatics, 2019, **44** (5) : 69-72.
7. Mao Yue, Song Xiao Yongoing, Jia Xiao Lin. Analysis about Parameters Selection Strategy of ECOM Solar Radiation Pressure Model for BeiDou Satellites[J]. Acta Geodaetica et Cartographica Sinica, 2017, **46** (11) : 1812-1821.
8. Liu Yu Xi, Jia Xiao Lin, Ruan Ren Gui. Beidou IGSO Satellite Orbit Determination Precision Analysis Based on New Attitude Control Mode[J]. Journal of Geodesy and Geodynamics, 2017, **37** (6) : 614-617.
9. Ji Guo Febng, Yang Zhi Qiang, Jia Xiao Lin, et al. A Study on the Impact of Yaw-Attitude and Solar Radiation Model on Precise Orbit Determination for Beidou Navigation Satellites[J]. Journal of Geodesy and Geodynamics, 2018, **38** (4) : 374-380.
10. Wang Chen, Guo Jing, Zhao Qi Le. Impact of Yaw Attitude on GPS/GLONASS Orbit and Clock Solutions[J]. Geomatics and Information Science of Wuhan University, 2017, **42** (5) : 624-629.
11. Zhou Pei Yuan, Du Lan, Fang Shan Chuan, et al. Analysis of Characteristics of QZSS Satellite Orbit and Clock Products during Yaw Attitude Model Switching[J]. Acta Geodaetica et Cartographica Sinica, 2016, **45** (3) : 274-281.
12. Lei Yongjun, Yuan Li. Angular momentum envelope analysis and angular momentum management of spacecraft[J]. China Space Science and Technology, 2017, **37** (6) : 1-9.

https://doi.org/10.1051/matecconf/202031604008

Progress of Electromagnetic Compatibility Design for Unmanned Aerial Vehicles

Yong Wu[1,2,a], Qishuang Ma[1] and Ping Xu[1]

[1]*College of Automation Science and Electrical Engineering, Beihang University ,Beijing 100191,China*

[2]*Beijing ELectro-Mechanical Engineering Institute, China Aerospace Science and Industry Corporation,Beijing 100074,China*

Abstract. The electromagnetic environment of UAV system was studied, and three key elements of electromagnetic interference(EMI) are analyzed. The research status of electromagnetic characteristics of UAV system at home and abroad was summarized from numerical simulation of EMI, experimental test of EMI and design of electromagnetic compatibility(EMC). The main contents include the research on electromagnetic characteristics of airborne antenna, EMI of data link and its mechanism, electromagnetic characteristics of internal cables, electromagnetic radiation test, electromagnetic conduction test, system-level EMC design, etc. The key technical challenges faced by the research on EMC of unmanned aerial vehicles at home and abroad are pointed out.

1 Research on Electromagnetic Environment

When UAV is flying, it is not only affected by electromagnetic environment caused by radiation of internal equipment, but also by external electromagnetic environment such as lightning, interference station, nuclear electromagnetic pulse and so on. If these interference can not be prevented reasonably, the performance of airborne equipment will be greatly affected[1].

Electromagnetic environment refers to the sum of power or field interference in different frequency range and the kinds and states of various interference sources when the UAV is performing its mission[2]. The external electromagnetic environment of UAV can be divided into natural electromagnetic environment and man-made electromagnetic environment. The internal electromagnetic environment is focused on here. The internal electromagnetic environment of UAV mainly includes antenna feed system, electromechanical system and circuit system.

2 EMI and EMC

EMI source, interference propagation path and sensitive equipment are called as three elements of EMI. Among them, EMI source refers to any component, equipment, system or natural phenomenon that produces interference, including intentional and unintentional emission; interference propagation means the medium that transmits EMI energy to sensitive equipment, which can be divided into conduction coupling and radiation coupling; sensitive equipment refers to the circuit and equipment that are susceptible to interference [3].

EMC refers to the electromagnetic environment in which the equipment or system meets the requirements and does not produce unbearable EMI to other equipment. According to the definition, EMC needs to satisfy three conditions: the electromagnetic disturbance generated by equipment or system in normal operation can not cause the performance degradation of other equipment; the electromagnetic disturbance generated by other equipment in electromagnetic environment has a certain tolerance; the electromagnetic disturbance that can not affect its performance degradation. Generally speaking, EMC is a subject that studies the coexistence and normal operation of various devices or systems in a specific working environment and limited spectrum resources.

3 Current Research Situation at Home and Abroad

3.1 Evaluation of EMC and its Criteria

In the 1970s, EMC testing technology was studied abroad. In 1996, the European Union began to implement the EMC license system for products. At present, developed countries such as Europe and the United States have made fruitful achievements in EMC evaluation criteria and standards. In the 1980s, China began to systematically study the national and industrial standards and specifications of EMC. At present, more than 30 national and military standards have been formulated.

[a] Corresponding author: 34846717@qq.com

The typical EMC evaluation system of UAV is shown in the evaluation criteria of UAV system EMC can be divided into two categories: system or equipment performance evaluation criteria based on deterministic electromagnetic environment characteristics and statistical analysis evaluation criteria based on stochastic environmental characteristics. Song Zuxun et al. [4] analyzed the evaluation criteria based on the characteristics of deterministic electromagnetic environment, but did not consider the EMI with obvious randomness. Zhu Zhiyu et al [5] gave the EMC standard of transient pulse group resistance. Li Bo et al. [6] obtained the magnetic field strength limit value of UAV system through the radiation and electromagnetic environment test of ground transmitter, and preliminarily formulated the EMC design evaluation criteria of UAV system. Zhang Yong et al [7] determined the evaluation index system of ship EMC, which can be used to guide the design of evaluation criteria of UAV EMC. Li Xiaojian et al. [8] preliminarily established system level EMC evaluation indexes and performance evaluation criteria.

EMC evaluation criteria is the basis of UAV avionics system design. According to the previous research experience, in the future, empirical analysis prediction method and index comprehensive construction method should be used in the model development to analyze and study the specific index, and optimize the whole system, so as to finally pass the consistency verification of calculation analysis and test.

3.2 Numerical Simulation of EMI of UAV

3.2.1 Study on Electromagnetic Characteristics of Airborne Antenna

Airborne antenna layout is one of the EMC design of UAV. Inter-antenna coupling, antenna pattern and near-field radiation have a great impact on the performance of airborne multi-antenna system, and are the main indicators to measure the layout of aircraft multi-antenna system. Among them, the coupling degree between antennas is the most important parameter to measure the EMC between antennas.

Tan Yuanyang[9] predicted the installation location of antenna. So the installed antenna not only meets the aerodynamic requirements of aircraft, but also achieves their main technical and tactical indicators and EMC between antennas. Yuan Jun[10] adopted adaptive fast multipole algorithm to solve the linear structure, analyzed the EMC pattern, isolation, interference power and other important indicators of airborne antenna system. Chen Jinji[11] proposed two methods for calculating antenna isolation. Eight airborne antennas were designed by HFSS simulation. Peng Tianhao[12] analyzed the EMI caused by the radio frequency excitation source in the meter-wave broadband plasma antenna and the overload of the instrument, the harmonic distortion of the measuring equipment and the harmonic noise in the communication frequency band. Liu Baihan [13] studied the numerical simulation technology of airborne antenna in radio frequency band on aircraft. Han Xuefeng[14] obtained the antenna pattern affected by the airframe and the layout design, and through the analysis of the electromagnetic radiation characteristics of the airborne antenna, obtained the optimum electromagnetic field distribution and antenna distribution area.

3.2.2 Research on EMI and its mechanism in data link

UAV airborne receiver is composed of multistage amplifier, mixer and filter. For UAV upstream data link, if the interference signal intensity entering the airborne receiver is large enough to exceed the dynamic range of various electronic devices in the receiver, the sensitivity of the receiver will be reduced and the signal-to-noise ratio will be reduced, which can be called blocking interference [15].

UAV communication is highly dependent on data link and vulnerable to battlefield electromagnetic environment during its mission. In UAV data link system, the receiver is sensitive to EMI[16]. EMI will cause the non-linear distortion of airborne receivers, which will lead to the interruption of data link communication and seriously threaten the battlefield survival of UAVs.

Jiao Yanwei[17] used cloud model method to study the comprehensive effectiveness of UAV in complex electromagnetic environment, and established the effectiveness evaluation index system of hierarchical structure of UAV in complex electromagnetic environment. Liu Ruiqi et al.[18] deduced the inequality of tracking interference hyperboloid and limited power interference range from the time domain and power condition of tracking interference, determined the effective interference range under different parameters, and proposed a reasonable jamming start-up strategy. Guo Hongjing[19] studied the cognitive strategy and cognitive process of UAV TT&C link interference. Zhang Xiaodong et al.[20] carried out the upstream main remote control lock-out sensitive frequency point injection method research, obtained the spectrum change rule of useful signal after interference, and obtained the lock-out mechanism through BPSK modulation and demodulation analysis. Guo Shuxia et al.[21] developed a scene view method based on GRA and the dynamic real-time driving method of microwave instrument according to the performance of real-time simulation and the intelligent function of simulation system, and simulated the dynamic interference of static radiation antenna in microwave anechoic chamber. Shen Yan'an [22] used Simulink simulation platform to build the UAV remote control link model based on direct spread spectrum technology. Song Zuxun et al. [23] calculated the interference probability or the number of received interference pulses, and compared with the fault-tolerant ability of the data chain.

3.2.3 Numerical simulation of electromagnetic characteristics of internal cables

Wire plays an important role in the EMC of UAV. Wires can be equivalent not only to receiving antennas, but also to transmitting antennas. According to statistics, the total weight of all kinds of cables on a large transport aircraft reaches several tons, and 60% of all kinds of EMI occurring on the aircraft is due to the electromagnetic coupling between cables.

Jiangdan[24] used three-dimensional electromagnetic simulation software CST to model and simulate the electromagnetic characteristics of cables inside the aircraft. Wang Chao[25] modeled and simulated the electromagnetic coupling characteristics of aircraft interconnected cables. Tan Yujian[26] studied the frequency domain solution method BLT equation and applied it to the analysis of transmission line crosstalk with arbitrary layout; also studied the time domain solution method-FDTD method, and deduced the iteration formula under lumped source loading and distributed source loading.

3.3 *EMI Test of Unmanned Aerial Vehicle*

3.3.1 Measurement of electromagnetic radiation

Radiation interference refers to the interference propagating in the form of electromagnetic waveform. The energy of this kind of interference is due to the radiation of the interference source, which propagates through the medium with the characteristics and regularity of electromagnetic waves. Huang Daqing et al. [27] analyzed and tailored the index of conduction and radiation test items in the National Military Standard. Song Huixuan et al. [28] established the transmission line model of the cable harness. The surface current distribution of the cable harness was calculated by using the multi-conductor transmission line model. Zhang Dongxiao et al. [29] took a certain type of unmanned target aircraft as the experimental object, built the test system by using the simulator construction method, put forward the radiation effect evaluation index and designed the dynamic monitoring system. Li Hongxiang [30] designed a test method to test the radiation disturbance and radio frequency magnetic field radiation immunity of UAV products, and built two test equipments.

3.3.2 Electromagnetic Conduction Test

Conducted interference can be divided into capacitance conduction coupling, resistance conduction coupling and inductance conduction coupling. Huang Daqing et al. [27] measured the conductive emission of the input power line of the tested equipment, aiming at the theoretical analysis and tailoring of the conductive emission CE101 of the power line, the theoretical analysis and tailoring of the conductive emission CE102 of the power line, and the theoretical analysis and tailoring of the conductive sensitivity CS101. Zhao Limin [31] deduced the INS heating model by system identification method, which provided a model basis for conduction interference analysis. Xu Ming et al. [32] conducted conduction sensitivity analysis and design on the basis of functional design. The improved power filter, RS232, PCB chip and cable design circuit or schematic diagram were given.

3.4 *EMC design of UAV system*

System-wide EMC design is to release EMC design indexes of equipment and subsystem according to the estimated electromagnetic environment. In recent years, some developed countries have gradually developed to the stage of EMC system design. The United States has the largest anechoic chamber, reverberation chamber and nuclear electromagnetic pulse test site in the world, as well as test equipment with high test accuracy, wide frequency band and high power. Rheinland T&V group has developed a large-scale industrial system-wide EMC testing equipment, which can be used for field testing of large-scale industrial machinery that can not be tested by ordinary EMC laboratories. TDK company of Japan built a large-scale high-performance anechoic chamber in Chiba in 2009, which can carry out EMC test, certification and rectification test of large-scale system [33].

At present, the consistency of system-wide EMC measurement in China is far less than that of equipment level EMC measurement, and the formulation of system-wide EMC test and corresponding standards is relatively lagging behind, so far there is no unified system-wide EMC test standard. The EMC laboratory of Beihang University has been able to undertake a variety of large-scale scientific research tasks, such as the design and evaluation of full aircraft EMC of special aircraft, the integration of EMC test system, EMC test, solutions to EMC problems and the design of Airborne special antenna [34]. First Academy of aerospace has the first laboratory in China that can carry out the system-wide EMC test capability of launch vehicle [35]. The EMC laboratory of AVIC helicopter design and Research Institute has passed the CNAs certification, and has the ability to undertake the EMC test of airborne equipment and subsystems, as well as the whole helicopter EMC test. Shanxi Haitai Electronic Co., Ltd. has developed a system-wide EMC automatic test system, which has the functions of electromagnetic environment measurement, system antenna coupling measurement, EMC safety margin measurement, power EMC characteristic test, etc.

4 Key technical issues

4.1 Consistency between heaven and earth in EMC

At present, most of the EMC characteristics of UAV systems are evaluated by matching test and comprehensive test. However, the test only examines the functions of wireless transceiver devices in different states of UAV, which can not meet the requirements of Heaven-Earth consistency instead of EMC test. In the course of flight, the electromagnetic environment of UAV is constantly changing. The equipment on the UAV is facing the electromagnetic environment impact caused by the high-power ground wireless equipment, and the electromagnetic environment caused by the wireless equipment in the aircraft and the target. Only by establishing an EMC test platform which is consistent with the actual flight environment, can it be possible to achieve the consistency between heaven and earth in the EMC test of UAV.

4.2 Evaluation and criteria of UAV electromagnetic compatibility

The most direct way to evaluate the electromagnetic compatibility of UAV system is the electromagnetic compatibility test. However, the test cycle is long and the cost is high, and the change of system equipment cannot be predicted. It is also difficult to carry out real-time inspection and evaluation of UAV electromagnetic environment in the flight process, resulting in the test data inconsistent with the actual situation. Therefore, it is necessary to develop reasonable EMC evaluation criteria to guide the design of UAV EMC.

4.3 UAV system level EMC design

EMC design of aircraft system includes data collection, model building, test verification, etc. in engineering development, it is often unable to carry out complete prediction and analysis due to schedule problems. After research and analysis, the author thinks that to carry out system level EMC design, we should first find out the weak link of UAV system design, and provide preliminary basis for EMC test; then combine comprehensive test to build the electromagnetic environment of UAV in flight; finally, carry out system EMC test to verify the coordination of wireless channel in dynamic electromagnetic environment Coordination and compatibility between man-machine and ground wireless equipment and electromagnetic compatibility of UAV system.

4.4 EMC Intelligent Design Technology

At present, the EMC analysis of UAV and other aircraft mainly depends on manual work. How to automatically realize the EMC test and evaluation needs the support of EMC expert system. The rapid development of artificial intelligence technology is widely used in robots, UAVs and other high-tech fields, but the EMC test system with artificial intelligence has not yet been reported. It is believed that EMC intelligent technology will be applied to UAV system design in the near future.

5 Summary

In recent years, the research on electromagnetic compatibility of UAV system at home and abroad has achieved fruitful results. However, the general evaluation criteria of UAV electromagnetic compatibility have not been formulated yet, and the existing criteria can not be used to guide the development of UAV models; Matching test and comprehensive test only examine the functions of wireless transceiver devices in different states of UAV, which can not meet the requirements of Heaven-Earth consistency instead of EMC test; the lack of a unified system level electromagnetic compatibility test standard in China makes it increasingly urgent to carry out relevant research work. With the increasing application of UAV, the research on electromagnetic compatibility of UAV system will attract more and more attention of aviation industry.

Now, deep research about antenna design and airborne antenna layout are carried out in our working group. In the future, we will do a lot of research on the anti-interference performance of data link system and the electromagnetic compatibility of UAV system.

References

1. Cai Rengang. EMC Principle, Design and Prediction Technology [M]. Beijing University of Aeronautics and Astronautics Press, 1997.
2. Gao Yougang. General Introduction to EMC [M]. Beijing University of Posts and Telecommunications Press, 2001.
3. Lu Yuzhong. Research on EMI Coupling and Protection of Small Aircraft in Complex Electromagnetic Environment [D], Xi'an University of Electronic Science and Technology, 2013.
4. Song Zuxun, Yu Bianzhang. Analysis of Electromagnetic Environment Effect of UAV System [J]. System Engineering and Electronic Technology, 2003; Volume 25, pp. 1108-1112.
5. Zhu Zhiyu, Zhang Bing, Liu Weiting. Application of Discrete Wavelet Filter in EMC Test of Aircraft [J]. Journal of Detection and Control. 2006; Volume 28, pp. 57-60.
6. Li Bo, Huang Daqing. Establishment of System Level Electromagnetic Radiation Index of UAV based on Electromagnetic Environment Effect [J]. Journal of Wuhan University (Science Edition). 2007; Volume 53, pp. 347-350.
7. Zhang Yong, Liu Junwei, Li Chao, etc. Research on Evaluation Index System and Application of Ship Electromagnetic Compatibility [J]. Ship Science and Technology. 2014; Volume 36, pp. 103-107.
8. Li Xiaojian, Kang Rui, Dai Fei, etc. Research on EMC Analysis Method based on Reliability Mathematics [J]. Telemetry and Telecontrol. 2010; Volume 31, pp. 66-70.
9. Tan Yuanyang. Analysis of scattering characteristics (cylindrical part) of aircraft antenna by UTD method [D]. Xi'an University of Electronic Science and Technology. 2004.
10. Yuan Jun. Research on high performance numerical algorithms for electromagnetic scattering and EMC analysis [D]. Xi'an University of Electronic Science and Technology. 2008.
11. Chen Jinji. Aircraft EMC prediction and simulation [D]. Xi'an University of Electronic Science and Technology. 2013.
12. Peng Tianhao. Research on the theory and key equipment of the METRE-WAVE broadband plasma antenna [M]. Shanghai Jiaotong University. 2015.
13. Liu Baihan. Research on numerical simulation technology for radio frequency compatibility of

aircraft [M]. Shenyang University of Aeronautics and Astronautics. 2017.
14. Han Xuefeng, Li Dong. Electromagnetic simulation analysis of airborne antenna based on finite difference time domain method [J]. Communication technology. 2017, 50 (11): 2472-2477.
15. Li Wei, Wei Guanghui, Pan Xiaodong et al. Prediction method of electromagnetic radiation effects of dual-frequency continuous wave in typical communication equipment band [J]. Systems engineering and electronic technology, 2016, 38 (11): 2474-2480
16. Guan Weiwei, Zhu Qianwang, Xu Jianzhong. Anti-jamming Analysis of UAV Data Link Based on SIR Equation [J]. Ship Electronic Engineering, 2013, 33 (1): 73-75
17. Jiao Yanwei, Hou Deting, Zhou Dongfang et al. Evaluation efficiency of UAV in complex electromagnetic environment [J]. Intensive laser and particle beam, 2014, 26 (7): 11-16
18. Liu Ruiqi, Yao Zhicheng, Yang Jian, etc. [J]. Study on Start-up Strategy of Tracking Jamming for UAV Remote Control Link [J]. Electro-optic and Control. 2018, 25 (4): 25-27.
19. Guo Hongjing. Research on interference perception and recognition technology of UAV TT&C link [D]. University of Electronic Science and Technology. 2014.
20. Zhang Xiaodong, Chen Ya, Xiao Xuerong. Research on EMI mechanism of main remote control data link of UAV [J]. Journal of Microwave Science. 2016, 32 (2): 90-96.
21. Guo Shuxia, Dong Zhongyao, Hu Zhantao, et al. Simulation of dynamic EMI environment for unmanned aerial vehicle data link[J]. EMC Design and its Application. 2013(8): 19-27.
22. Shen Yanan. Simulation study on electromagnetic environment adaptability of UAV remote control data link [J]. Value Engineering. 2012 (5): 178-179.
23. Song Zuxun, Yu Bianzhang. Adaptability of UAV data link to radar EMI [J]. Systems Engineering and Electronic Technology. 2004, 26 (12): 1790-1793.
24. Jiangdan. Research on electromagnetic conduction and radiation characteristics and suppression methods of multi-electric aircraft system [D]. Nanjing University of Aeronautics and Astronautics, 2016.
25. Wang Chao. EMI and Airworthiness of aircraft EWIS wiring [D]. Civil Aviation University of China. 2012.
26. Tan Yujian. Research on EMC analysis method for complex systems [D]. National University of Defense Science and Technology. 2009.
27. Huang Daqing, Li Bo. EMC Conduction and Radiation Test Project Cutting Method for UAV [J]. Optical Precision Engineering. 2009, 17 (2): 380-387.
28. Song Huixuan, Wu Gang, Zhu Long, etc. EMC analysis of radiation from cables and cables harness inside aircraft [R]. National Conference on Microwave and Millimeter Wave. 2013: 1532-1535.
29. Zhang Dongxiao, Chen Ya, Cheng Erwei, et al. Experimental study on continuous wave electromagnetic radiation effects of an unmanned target aircraft [J]. Journal of Hebei Normal University (Natural Science Edition). 2017, 41 (1): 39-44.
30. Li Hongxiang. Testing of Radiation Disturbance and Radiation Immunity of UAV [D]. Heilongjiang University. 2017.
31. Zhao Limin. Research on Conducted EMI Analysis Method for Aircraft Electrical System [D]. Harbin Engineering University. 2012.
32. Xu Ming, Chen Yan. Conduction sensitivity analysis and design of an airborne flight controller [J]. Ship electronics engineering. 2016.
33. Zhu Chengbang, Hua Yang. Ship Technology Process Elements and Electromagnetic Compatibility Test[J]. Ship Electronic Engineering, 2013; Volume 33, pp. 171-173.
34. Sun Guangjing. Progress the development of ship EMC Technology [J]. Ship Electronic Engineering. 2007; Volume 27, pp. 20-22, pp. 40.
35. Lai Shunxiang, Yang Huiqin. Discussion on EMC Test Method of Space Launch Site [J]. Ship Electronic Engineering. 2004; Volume 24, pp. 160-162.

Research on Cooperative Strategy of Multiple Rockets

Ke Yi[1], Si-hong She[1], Shun Zhang[1], Bang-ya Hu[2,a], Meng Liu[2]

[1]*Systems Engineering Research Institute, Beijing, China*
[2]*School of Aerospace Engineering, Beijing Institute of Technology, Beijing, China*

Abstract. In order to enhance the penetration ability of the rocket and the damage degree of the target, the trajectory design and optimization of the rockets are carried out, and the cooperative strategy of multiple rockets is designed. Firstly, the particle motion model of rocket is established, and the power and guidance scheme are given. Then, a cooperative attack strategy based on different launch points for multiple rockets is proposed, which transforms the cooperative attack into the optimization of launch points. The optimization model of the launch point under the expected attack time is established, and the model is solved by the cultural algorithm. The distribution of the launch point under different attack time is obtained and fitted as curves under corresponding conditions. Finally, aiming at the problem of multiple rockets cooperative combat, using the obtained curves, three kinds of cooperative strategies which are suitable for different attack time requirements are proposed and verified by simulation.
Keywords. multiple rockets, cooperative combat, trajectory design and optimization, cultural algorithm

1 Introduction

With the development of anti-missile technology, more and more attention has been paid to multiple missiles cooperative attack and defense, and cooperative guidance technology has also been developed rapidly. If multiple rockets can attack the target together, the penetration ability and damage degree of the rocket can be greatly enhanced. Therefore, this paper proposes a scheme to realize the cooperative combat of multiple rockets by changing the launching point without changing the guidance parameters of rockets.

At present, scholars have done some research on missile cooperative combat. In [1] it is determined that the MRSI (multiple rounds simultaneous impact) scheme is to find out all possible trajectories for a certain range, and then select the trajectories satisfying MRSI. Reference [2] made some researches on MRSI of one-dimensional correction missiles launched by naval gun, put forward a fire control algorithm to achieve MRSI by changing the launching angle and damping ring opening time, and then in [3] it is improved on the basis of the algorithm to meet the needs of fast and violent attack on the target. In order to improve the trajectory calculation accuracy of MRSI of single gun, a local interpolation algorithm is proposed based on the two-dimensional particle trajectory equation and the two-dimensional vacuum trajectory oblique range formula in [4]. Although some algorithms have been proposed by some scholars for MRSI, most of the existing researches are still on the guided shell of naval gun, and the algorithm still has the problems of long calculation time and large calculation amount. A guidance problem for impact time control applicable to salvo attacks is considered based on the sliding mode control in [5]. Time constraints are achieved by designing an impact-time vector guidance law in [6]. Reference [7] suggested a "Close-Formation Approach" attack tactics as a new tactics to operate multiple anti-ship missiles. However, because the rockets generally does not have the ability of inter missile communication, the existing cooperative formation technology cannot be applied to the cooperative combat of rockets. At present, there is no mature and effective strategy for the research of rockets cooperative combat.

In this paper, a three degree of freedom particle trajectory model of rocket is established, and the power and guidance scheme are designed. Aiming at the cooperative combat of multiple rockets, the cooperative attack is realized by changing the launch site of rockets, and the cooperative attack is transformed into the optimization of launch point. The optimization model of cooperative attack is established. The distribution curve of launch point with the same attack time is obtained by solving the optimization model. Three cooperative strategies are proposed and verified by simulation.

2 Modeling of rocket motion

2.1 Rocket motion model

The equations of motion of the rocket's center of mass in the launch coordinate system are

[a] Corresponding author: 644482455@qq.com

$$\begin{cases} m\frac{dV}{dt} = P\cos\alpha\cos\beta - X - mg\sin\theta \\ mV\frac{d\theta}{dt} = P(\sin\alpha\cos\gamma_V + \cos\alpha\sin\beta\sin\gamma_V) \\ +Y\cos\gamma_V - Z\sin\gamma_V - mg\cos\theta \\ -mV\cos\theta\frac{d\psi_V}{dt} = P(\sin\alpha\sin\gamma_V \\ -\cos\alpha\sin\beta\cos\gamma_V) + Y\sin\gamma_V + Z\cos\gamma_V \\ \frac{dx}{dt} = V\cos\theta\cos\psi_V \\ \frac{dy}{dt} = V\sin\theta \\ \frac{dz}{dt} = -V\cos\theta\sin\psi_V \\ \frac{dm}{dt} = -m_c \\ \alpha = -\frac{m_z^{\delta_z}}{m_z^{\alpha}}\delta_z \\ \beta = -\frac{m_y^{\delta_y}}{m_y^{\beta}}\delta_y \\ \varepsilon_1 = 0 \\ \varepsilon_2 = 0 \\ \varepsilon_3 = 0 \\ \varepsilon_4 = 0 \end{cases} \tag{1}$$

where, t is time, m is rocket mass, V is rocket velocity, θ is flight path angle, ψ_V is flight path deflection angle, P is thrust, X is drag, Y is lift, Z is lateral force, α is angle of attack, β is angle of sideslip, γ_V is slop angle of velocity, x, y, z are position coordinates, m_c is the mass variation, g is the acceleration of gravity, δ_y, δ_z are rudder deflection angle and elevator angle, $m_y^{\delta_y}, m_z^{\delta_z}, m_y^{\beta}, m_z^{\alpha}$ are rudder control efficiency, elevator control efficiency, yaw static stability derivative, pitch static stability derivative, $\varepsilon_1, \varepsilon_2, \varepsilon_3, \varepsilon_4$ are control equations.

2.2 Power and guidance scheme

The rocket adopts solid rocket motor, and the thrust scheme is

$$P = \begin{cases} P_1 & 0 < t \le t_1 \\ P_2 & t_1 < t \le t_2 \\ 0 & t > t_2 \end{cases} \tag{2}$$

In [8] a biased proportional guidance law with terminal angular constraint was given

$$a_{yc} = KV_m\dot{q}_z + V_m\frac{\theta_f - \theta}{t_{go}} - KV_m\frac{\theta_f - q_z}{t_{go}} \tag{3}$$

where, K is proportional Coefficient, time to go $t_{go} = r / V_m$, θ_f is flight path angle at hit, q_z is line of sight angle.

The longitudinal flight program of the rocket is: the initial flight path angle is θ_0, when $0 < t \le t_1$, the rocket is uncontrolled; when $t_1 < t \le t_2$, the rocket climbs according to the pitching angle command ϑ_c ;when $t_2 < t \le t_3$, the rocket is uncontrolled; when $t > t_3$ and the missile-target distance $r \ge r_e$, the rocket glides at a fixed angle of attack, and the angle of attack command is α_c; when $r < r_e$, the rocket uses the proportional guidance law shown in (3). The lateral flight program of the rocket is: when $y \ge y_c$, the rocket flies without control; when $y < y_c$, the rocket uses proportional guidance law to fly to the target, and the guidance instruction is $a_{zc} = K_z V_m \dot{q}_y$, where K_z is proportional guidance coefficient. The inclined channel flight program of the rocket is: when the rocket is launched, the slope angle command is γ_c.

3 Optimization model and solution of multiple rockets cooperative attack

3.1 Optimization model of cooperative attack

In order to achieve the cooperative attack of multiple rockets, different launch points are selected to control the attack time of rockets, and the cooperative attack problem is transformed into the distribution of launch points with a given attack time. By solving the optimization model, the distribution of launch points under the same attack time is obtained, and the distribution curve of launch points under the given attack time is fitted. If the attack time interval of the rockets is given, the appropriate launch position can be selected through the distribution curve to achieve cooperative attack.

Let the coordinates of the rocket launch point in the ground coordinate system be (x_0, y_0, z_0), where $y_0 = 0$, and take x_0, z_0 as the design variable, then the expression of the design variable is

$$\boldsymbol{X} = [x_1, x_2]^T = [x_0, z_0] \tag{4}$$

If the attack time of rocket launched by (x_0, y_0, z_0) is t, the objective function of the minimum attack time optimization model is

$$\min F_1(X) = t \tag{5}$$

The objective function of the maximum attack time optimization model is

$$\max F_2(X) = t \tag{6}$$

The objective function of the optimization model of rocket attacking target with expected attack time $\bar{t}$ is

$$\min F_3(X) = |t - \bar{t}| \tag{7}$$

Equations (5) to (7) are solved several times under the constraint conditions, and a set of coordinates of the launch point under the corresponding conditions are obtained. The distribution curve of the launch point can be obtained by fitting the coordinate points.

The constraints of design variables are

$$\begin{cases} X_{\min} = [x_{0\min}, z_{0\min}] = [-10000m, -10000m] \\ X_{\max} = [x_{0\max}, z_{0\max}] = [10000m, 10000m] \end{cases} \tag{8}$$

Therefore, the feasible region of the optimization model is defined as

$$\Omega = [X_{\min}, X_{\max}] \tag{9}$$

The angle between velocity vector and longitudinal axis of the body is defined as the actual angle of attack of rocket in three-dimensional space

$$\chi = \arccos(\cos\alpha\cos\beta) \tag{10}$$

The process and terminal constraints that need to be satisfied in the course of rocket flight are

$$g_i(X) \le 0, i = 1, 2, 3, \cdots, 6 \tag{11}$$

where

$$\begin{cases} g_1(X) = |\chi| - \chi_m \\ g_2(X) = \theta_{fc} - |\theta_f| \\ g_3(X) = |a_y| - a_{ym} \\ g_4(X) = |a_z| - a_{zm} \\ g_5(X) = V_{\min} - V_f \\ g_6(X) = V_f - V_{\max} \end{cases} \tag{12}$$

where, χ_m, a_{ym}, a_{zm} are maximum values of corresponding parameters, θ_{fc} is desired falling angle, $V_{\min}, V_{\max}$ are minimum and maximum velocity.

The optimal mathematical models of the minimum attack time, the maximum attack time and the expected attack time of the rockets are shown in (13) to (15).

$$\begin{aligned} \min\ & F_1 = F_1(X) \\ \text{s.t.}\ & g_i(X) \le 0, i = 1, 2, 3, \cdots, 6 \\ & X \in \Omega \end{aligned} \tag{13}$$

$$\begin{aligned} \max\ & F_2 = F_2(X) \\ \text{s.t.}\ & g_i(X) \le 0, i = 1, 2, 3, \cdots, 6 \\ & X \in \Omega \end{aligned} \tag{14}$$

$$\begin{aligned} \min\ & F_3 = F_3(X) \\ \text{s.t.}\ & g_i(X) \le 0, i = 1, 2, 3, \cdots, 6 \\ & X \in \Omega \end{aligned} \tag{15}$$

3.2 Optimization model solution based on cultural algorithm

Cultural algorithm is a two-level evolutionary system based on knowledge, which contains two evolutionary spaces, one is belief space composed of experience and knowledge acquired in the process of evolution, the other is population space composed of specific individuals, through evolutionary operation and performance evaluation to carry out its own iterative solution[9,10]. Cultural algorithm has been widely used in the field of multi-objective optimization. The framework and flow chart of cultural algorithm are shown in Fig. 1 and Fig. 2 respectively.

The flow of cultural algorithm to solve the problem of optimization of rocket attack time is as follows:

(1) Initialize the population space: Within the range of design variables, the initial population $\mathbf{P}_0$ with the scale of n is randomly generated, i.e. n groups of rocket launch point coordinates. Then, the n groups of coordinates generated are respectively substituted into the rocket's trajectory model to obtain the attack time of the rocket. The fitness value $\mathbf{F}$ is calculated according to the target function.

(2) Initialize the knowledge space: The optimal value in population $\mathbf{P}_0$ is selected as the initial value of situation knowledge; the lower limit $\mathbf{l}_0$ and the upper limit $\mathbf{u}_0$ are set as $\mathbf{X}_{\min}$ and $\mathbf{X}_{\max}$ in (8) according to the upper and lower limits of design variables, and the evaluation value $\mathbf{L}_0$ corresponding to the lower limit $\mathbf{l}_0$ and the evaluation value $\mathbf{U}_0$ corresponding to the upper limit $\mathbf{u}_0$ are initialized as $+\infty$.

(3) Update the population space: According to the cultural algorithm, the iteration step and evolution direction of the design variables of the $t+1$ generation are determined, and n new launch point coordinates are generated, that is, the new group $\mathbf{P}_{t+1}$. The corresponding attack time is obtained by substituting n new launch point coordinates into the rocket trajectory model, and the fitness value of the individual population is calculated according to the objective function of the optimization model.

(4) Update the knowledge space: According to the cultural algorithm to update the situation knowledge of the knowledge space, the optimal solution s^{t+1} of the first $t+1$ generation is obtained; the normative knowledge of the knowledge space is updated, that is, the upper limit l and the lower limit $\mathbf{u}$ of the design variables and the corresponding evaluation values $\mathbf{L}$ and $\mathbf{U}$.

(5) Judge whether the termination conditions are met. If not, go back to step (3).

(6) At the end, the optimal solution is output.

According to the optimization model established in Sec 3.1, the coordinates of launch point in launch coordinate system under different attack time can be obtained by using the above methods.

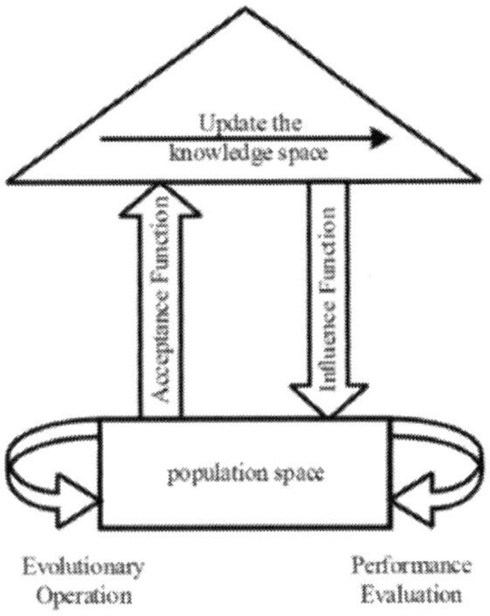

Figure 1. Basic framework of cultural algorithm.

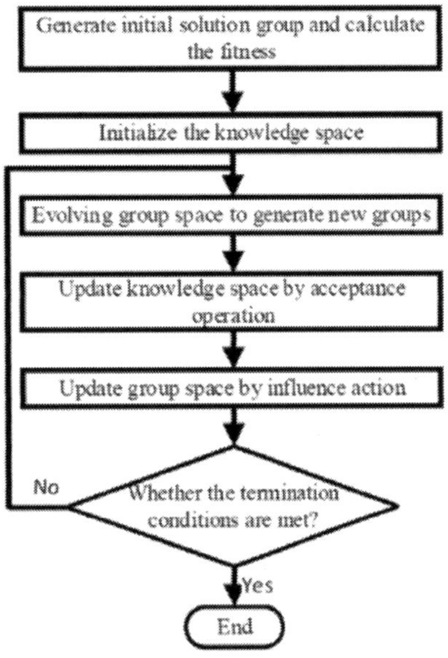

Figure 2. Flow chart of cultural algorithm.

4 Cooperative strategy of multiple rockets

In order to achieve multiple rockets attacking the target in a certain order, it can be achieved by setting the position of the rocket launching unit. The specific process is as follows:

(1) According to the solution method given in Sec. 3.2, m sets of launch point coordinates corresponding to the minimum attack time $t_{\min}$ and m sets of launch point coordinates corresponding to the maximum attack time $t_{\max}$ are obtained respectively.

(2) In the rocket attack time range $[t_{\min}, t_{\max}]$, the expected attack time of the rocket is $\overline{t}_i = t_{\min} + i\Delta t, i = 1, 2, \cdots, n$. Using the same solution method, the coordinates of m launch points corresponding to the expected attack time $\overline{t}_i$ are obtained respectively.

(3) Let $\boldsymbol{\overline{T}} = [t_{\min}, \overline{t}_1, \overline{t}_2, \cdots, \overline{t}_n, t_{\max}]$, use the least square method to fit the coordinates of m group launch points corresponding to each attack time in $\overline{T}$. The distribution curve of launch points under attack time T_i is recorded as S_i, and its expression is recorded as $x_i = a_i z_i^2 + b_i z_i + c_i, z_i \in [z_{i\min}, z_{i\max}]$.

(4) Select the launch point coordinates according to the actual combat tasks, and consider the cooperative attack of w rockets on a fixed target. The selection methods of launch points for different cooperative strategies are as follows:

Suppose that multiple rockets attack the target at the same time, as shown in Fig. 3, the cooperative strategy is

a. If the attack time T_e is equal to the attack time $\overline{T}_j$ corresponding to a certain launch point distribution curve S_j in (3): Select S_j as the launch point distribution curve S_e; if the attack time T_e is not equal to any element in $\boldsymbol{\overline{T}}$, then find the attack time T_k in $\boldsymbol{\overline{T}}$ to make it meet $T_k < T_e < T_{k+1}$, and the analytical expression of the launch point distribution curve S_e corresponding to the attack time T_e is

$$x_e = x_{k+1} + \frac{T_{k+1} - T_e}{\Delta t}(x_k - x_{k+1}) \tag{16}$$

The analytical formula of S_e can be sorted as

$$\begin{gathered} x_e = a_e z_e^2 + b_e z_e + c_e \\ z_e \in [\max(z_{k\min}, z_{k+1\max}), \min(z_{k\min}, z_{k+1\max})] \end{gathered} \tag{17}$$

where,

$$\begin{cases} a_e = a_{k+1} + \dfrac{(T_{k+1} - T_e)(a_k - a_{k+1})}{\Delta t} \\ b_e = b_{k+1} + \dfrac{(T_{k+1} - T_e)(b_k - b_{k+1})}{\Delta t} \\ c_e = c_{k+1} + \dfrac{(T_{k+1} - T_e)(c_k - c_{k+1})}{\Delta t} \end{cases} \tag{18}$$

The curve S_e domain is divided into w-1 equal parts, so that the z coordinate of the launch point of the rocket is taken as the point of w-1 section of the domain, and then the coordinates of the launch point are determined according to the fitting formula.

Suppose that multiple rockets attack the target in order, as shown in Fig. 4, the cooperative strategy is

b. If the attack time $T_e^1, T_e^2, \cdots, T_e^w$ of multiple rockets attacking the target in sequence is equal to $\overline{T}_j, \overline{T}_{j+1}, \cdots, \overline{T}_{j+w-1}$ in $\boldsymbol{\overline{T}}$: the launching points of w rockets $\{M_1, M_2, \cdots, M_w\}$ are respectively located at $S_j, S_{j+1}, \cdots, S_{j+w-1}$, where M_1, M_w is respectively located at the endpoint, and the launching point of the ith rocket M_i is located at the *i-(w-1)*th equinox of curve S_{j+i-1}. Among them, the point of *w*-1 section of curve S_{j+i-1} are sorted from small to large according to z coordinate. Only the situation that the attack interval is Δt is given here, if the expected attack time difference is an integral multiple of Δt, the principle is the same and will not be repeated.

c. If the attack time of multiple rockets attacking the target in sequence is arbitrary $T_e^1, T_e^2, \cdots, T_e^w$: Firstly, according to the strategy *a*, obtain the distribution curve S_e^1 and S_e^w of the launch points whose attack time is T_e^1 and T_e^w respectively; then connect the diagonal ends of S_e^1 and S_e^w to obtain the straight line *S* (as shown in Fig. 4); finally, divide the straight line *S* into *w-1* equal parts, and select the equal diversion points as the launch points of other rockets. If the given attack time difference is not equal, we can take a point on the straight line *S* according to the proportion of attack time difference.

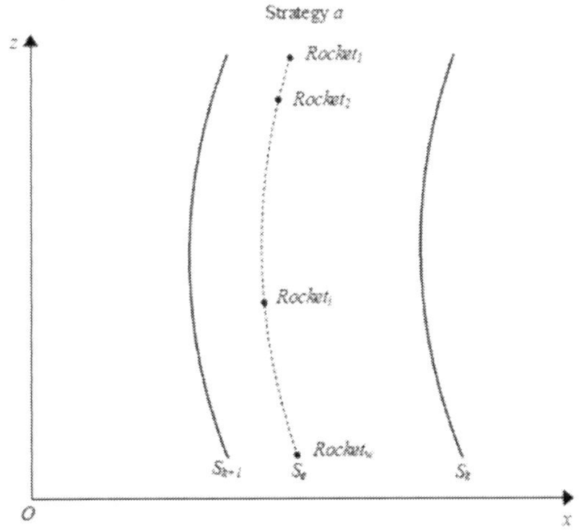

Figure 3. Cooperative strategy for multiple rockets attacking targets at the same time.

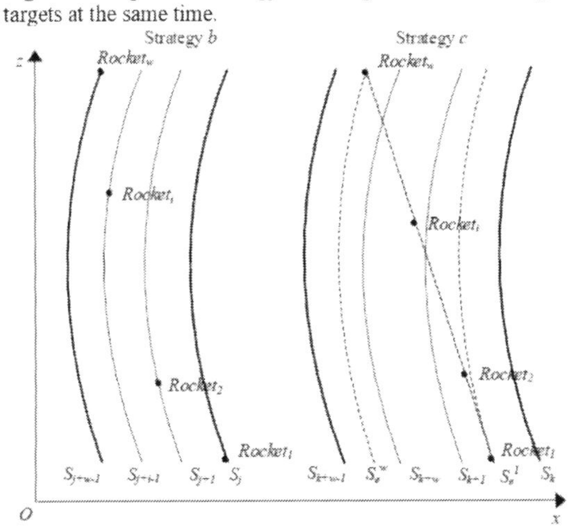

Figure 4. Cooperative strategy of multiple rockets attacking targets in sequence.

5 Simulation results and analysis

Based on the rocket motion model established in Sec. 2.1 and the power and guidance scheme in Sec. 2.2, the minimum and maximum attack time optimization models are solved by using the method in Sec. 3.2. See Table 1 for simulation parameter values.

Table 1. Simulation parameter values.

Variable name	Value
t_1	5s
t_2	22s
t_3	200s
P_1	120kN
P_2	55kN
θ_0	50°
ϑ_c	45°
r_e	30km
α_c	8°
y_c	20km
γ_c	0
K	5
K_z	3
χ_m	10°
a_{ym}	10°
a_{zm}	10°
θ_{fc}	80°
$V_{\min}$	600m/s
$V_{\max}$	900m/s

The population size is set to 20, and the termination generation of the algorithm is set to 100. The minimum attack time of the rockets is 280s, and the maximum attack time is 302.5s. Set the expected attack time $\bar{T} = 282, 284, 286, \cdots, 300$, calculate 50 times for each optimized model of the expected attack time, and get the distribution of launch points as shown in Fig. 5.

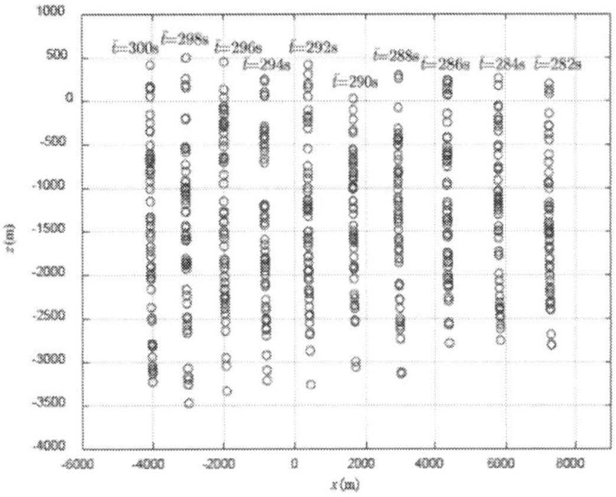

Figure 5. Distribution of launch points under different expected attack times.

The distribution curves of the launching points are fitted, as shown in Table 2.

It can be seen from Table 2 that when the least square method is used for the quadratic polynomial fitting of the optimization results, the complex correlation coefficient approaches to 1, indicating that the fitting method has high precision, and the fitting results are shown in Fig. 6. To verify the fitting results, select 100 points evenly on each curve for simulation, and the attack time error is shown in Fig. 7.

In order to verify the multiple missiles cooperative strategy proposed in Sec. 4, four rockets are assumed to attack the target cooperatively, and three cooperative strategies proposed in this paper are used for simulation.

1) Four rockets are required to hit the target at the same time, and the attack time is 291s. According to strategy *a*, the distribution curve of launch point with attack time of 291s is deduced as follows

$$x_{e291} = 14.81 z_{e291}^2 + 25.37 z_{e291} + 1000 \quad (19)$$
$$z_{e291} \in [-3.061\text{km}, 0.022\text{km}]$$

The launching positions of the four rockets are shown in Fig. 8 (marked with "○" in the figure), and the simulation results are shown in Fig. 9 to Fig. 11.

It can be seen from Fig. 8 to Fig. 11 that the four rockets meet the end velocity constraint, and the time of convergence to 30m is 290.98s, 290.99s, 290.98s, 290.99s, and the maximum attack time difference is 0.01s.

2) Four rockets are required to hit the target in a certain order, and the attack time is 290s, 292s, 294s and 296s respectively. According to strategy *b*, select the launch location as shown in Fig. 12 (marked with "○" in the figure), and the simulation results are shown in Fig. 13 to Fig. 15.

Table 2. Fitting results of distribution curve of emission points.

Attack time(s)	Fitting formula(m)	Range of independent variable	Complex correlation coefficient (R^2)
280	$x_1 = 8.72z_1^2 + 3.79z_1 + 8816$	[-2.499km,0.005km]	0.9189
282	$x_2 = 18.41z_2^2 + 31.35z_2 + 7274$	[-2.806km,0.193km]	0.9830
284	$x_3 = 17.98z_3^2 + 30.26z_3 + 5798$	[-2.756km,0.258km]	0.9831
286	$x_4 = 15.97z_4^2 + 26.77z_4 + 4366$	[-2.789km,0.232km]	0.9798
288	$x_5 = 15.54z_5^2 + 25.74z_5 + 2978$	[-3.139km,0.291km]	0.9893
290	$x_6 = 15.90z_6^2 + 28.32z_6 + 1640$	[-3.061km,0.022km]	0.9883
292	$x_7 = 13.71z_7^2 + 22.41z_7 + 360$	[-3.264km,0.412km]	0.9798
294	$x_8 = 12.98z_8^2 + 21.24z_8 - 853$	[-3.215km,0.245km]	0.9906
296	$x_9 = 12.90z_9^2 + 21.72z_9 - 1994$	[-3.337km,0.447km]	0.9906
298	$x_{10} = 11.60z_{10}^2 + 18.38z_{10} - 3065$	[-3.478km,0.498km]	0.9937
300	$x_{11} = 10.98z_{11}^2 + 16.73z_{11} - 4072$	[-3.229km,0.417km]	0.9919
302.5	$x_{12} = 25.35z_{12}^2 + 70.08z_{12} - 5192$	[-3.414km,-0.122km]	0.9283

https://doi.org/10.1051/matecconf/202031604009

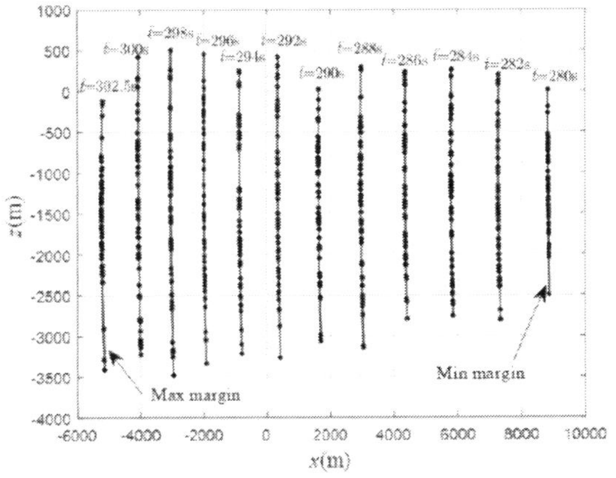

Figure 6. Distribution curve of launch points with different attack time.

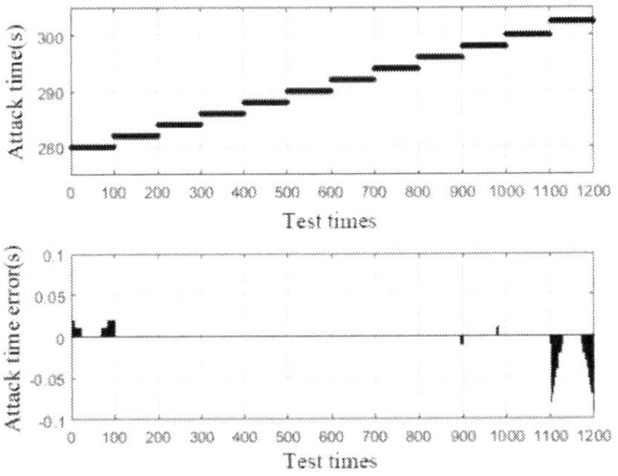

Figure 7. Attack time error.

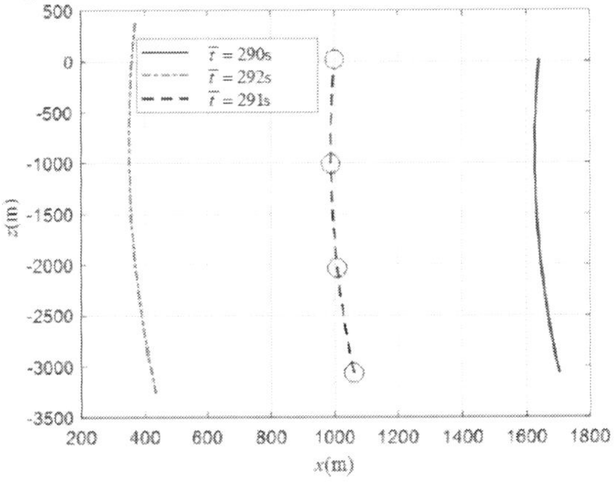

Figure 8. Launch point location of cooperative strategy *a*.

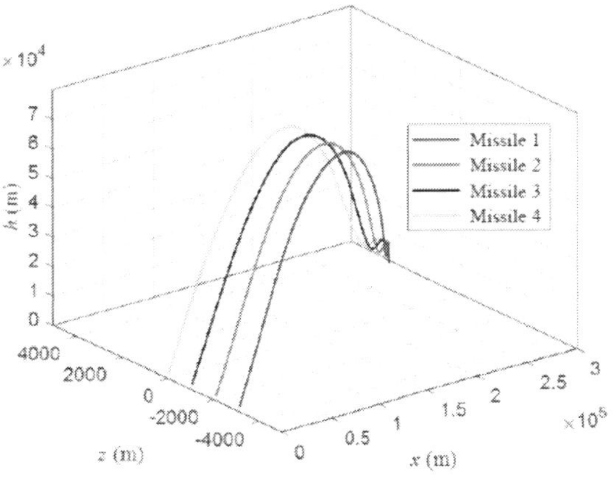

Figure 9. Three dimensional trajectory diagram of cooperative strategy *a*.

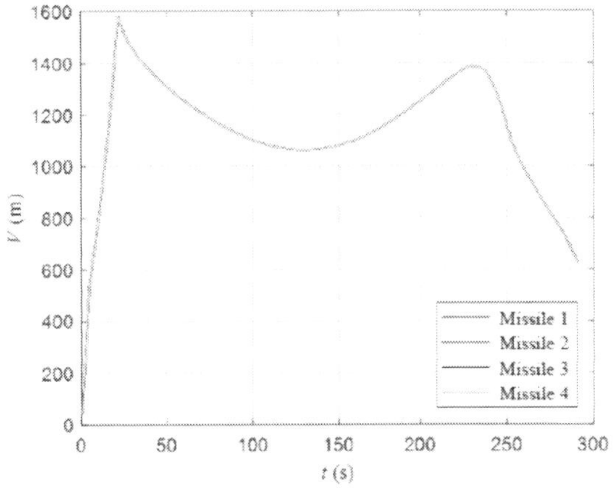

Figure 10. Speed curve of collaborative strategy *a*.

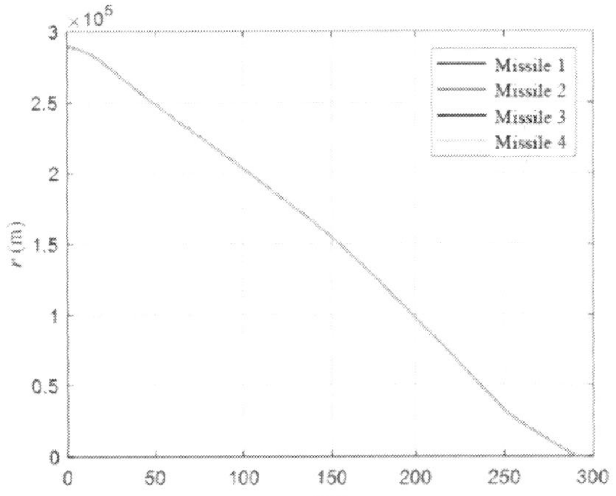

Figure 11. Missile target distance curve of cooperative strategy *a*.

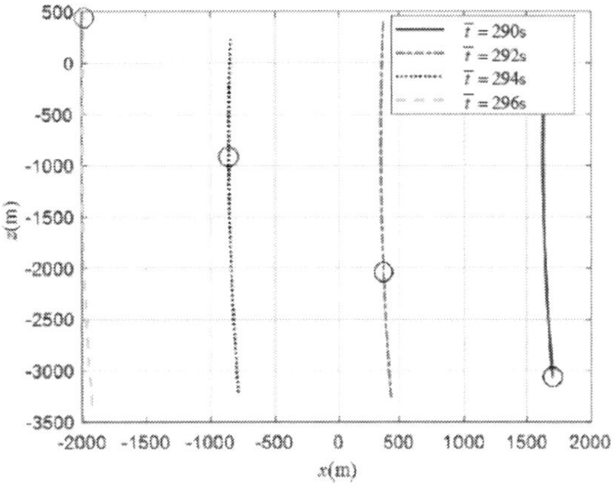

Figure 12. Launch point location of cooperative strategy *b*.

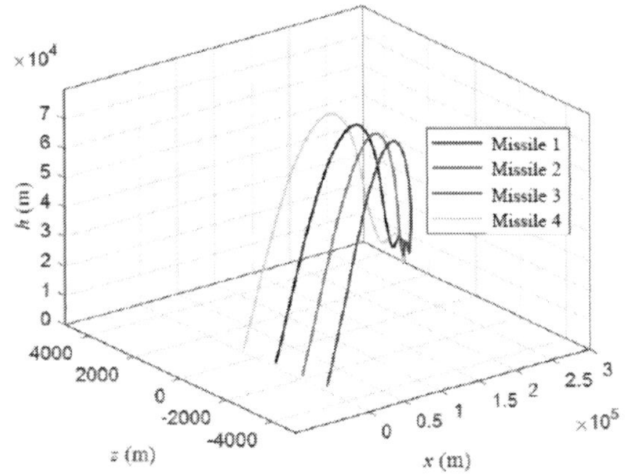

Figure 13. Three dimensional trajectory diagram of cooperative strategy *b*.

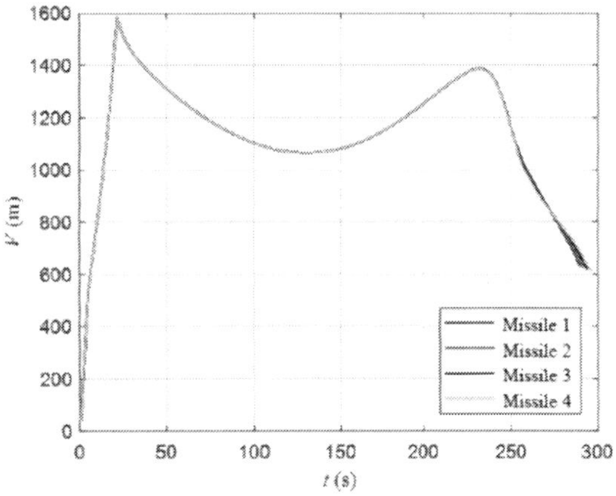

Figure 14. Speed curve of collaborative strategy *b*.

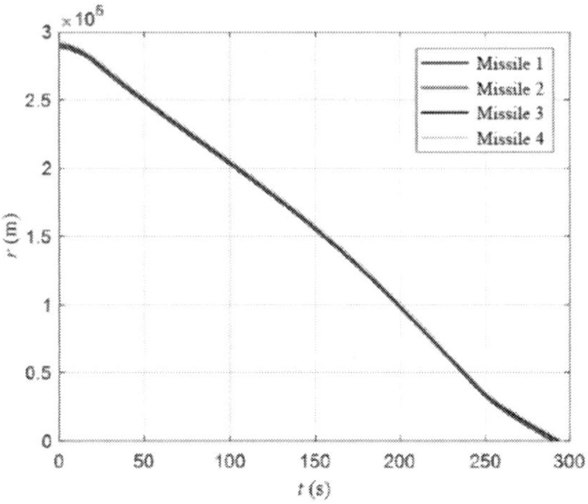

Figure 15. Missile target distance curve of cooperative strategy *b*.

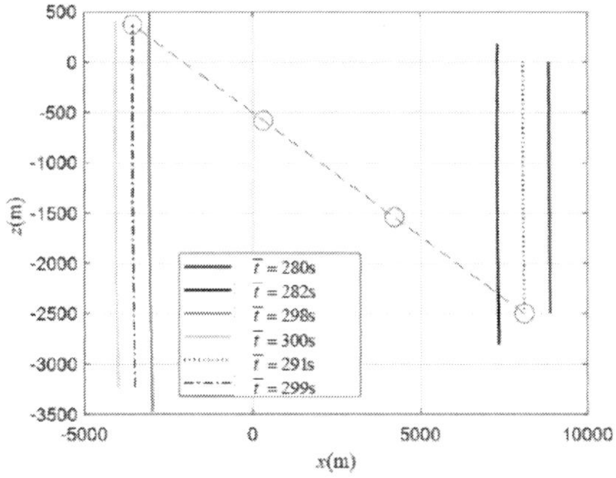

Figure 16. Launch point location of cooperative strategy *c*.

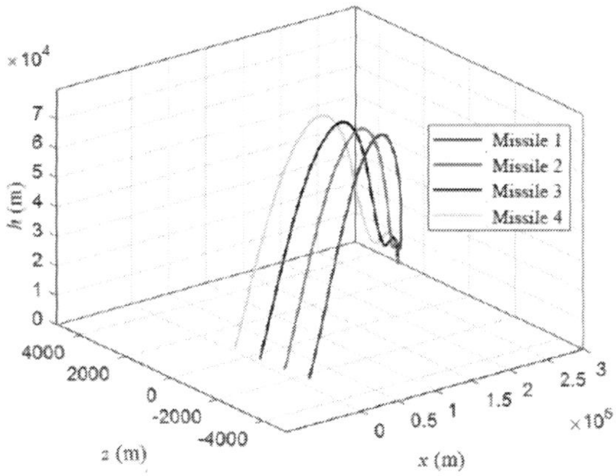

Figure 17. Three dimensional trajectory diagram of cooperative strategy *c*.

 https://doi.org/10.1051/matecconf/202031604009

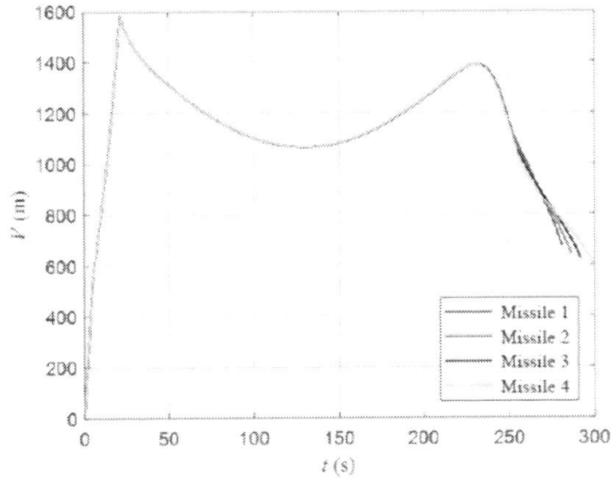

Figure 18. Speed curve of collaborative strategy *c*.

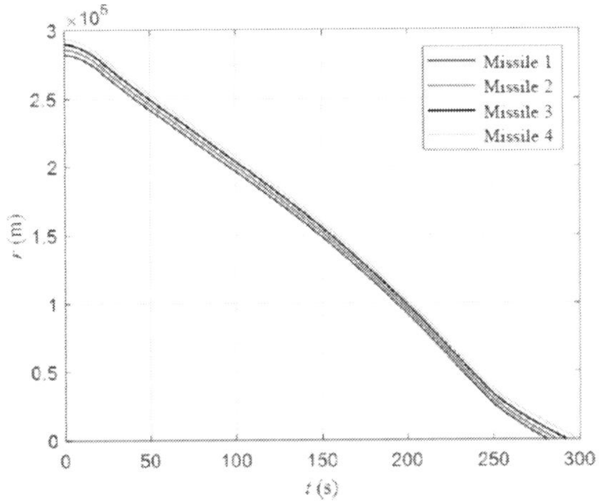

Figure 19. Missile target distance curve of cooperative strategy *c*.

It can be seen from Fig. 12 to Fig. 15 that the four rockets meet the end velocity constraint, and the time of convergence to 30m is 290s, 292s, 294s and 295.99s.

3) four rockets are required to hit the target at 6s interval, and the attack time is 281s, 287s, 293s and 299s respectively. According to strategy *c*, the distribution curve of launch points with attack time of 281s and 299s is deduced as follows

$$\begin{aligned} x_{e281} &= 13.57z_{e281}^2 + 17.57z_{e281} + 8045 \\ z_{e281} &\in [-2.499\text{km}, 0.005\text{km}] \end{aligned} \quad (20)$$

$$\begin{aligned} x_{e299} &= 11.29z_{e299}^2 + 17.56z_{e299} - 3569 \\ z_{e299} &\in [-3.229\text{km}, 0.417\text{km}] \end{aligned} \quad (21)$$

recorded as S_{281} and S_{299} respectively. The launch positions of the four rockets are shown in Fig. 16 (marked with "○" in the figure), and the simulation results are shown in Fig. 17 to Fig. 19.

It can be seen from Fig. 16 to Fig. 19 that the four rockets meet the end velocity constraint. The time of convergence to 30m is 280.98s, 286.22s, 292.05s and 298.98s. The attack time interval is 5.24s, 5.83s and 6.93s, respectively. Compared with the expected 6s interval, the error is within 1s.

6 Conclusions

In this paper, the trajectory design and optimization of the rockets are studied, the motion model of the rocket's center of mass is established, and the power and guidance scheme are designed. In order to solve the problem of multiple rockets cooperative attack, a multiple rockets cooperative attack scheme based on different launch points is proposed. The cooperative attack problem of rockets is transformed into the optimization problem of launch point. The optimization problem is solved by cultural algorithm, and the distribution curve of launch point with different attack time is fitted. Based on the distribution curve of launch points, the cooperative combat strategy of multiple rockets is proposed to attack targets simultaneously or in a specified order. The next problems need to be solved are the elastic vibration, the coupling between the aerodynamic force and the control system, and the aerodynamic heat of the rocket when it enters the atmosphere at a high speed from the highest point of the trajectory.

References

1. L. Y. Xie, R. Liao, X. Q. Wang, Command Control & Simulation, **4**, 30 (2009)
2. Y. Huang, D. H. Wang, H. C. Wang, J. Hu, W. H. Xie, Journal of Projectiles, Rockets, Missiles and Guidance, **5**, 127 (2012)
3. Y. Huang, J. D. Huang, J. You, Journal of Projectiles, Rockets, Missiles and Guidance, **5**, 6 (2018)
4. L. D. Cheng, Y. Jiang, Y. C. Niu, Journal of Ballistics, **2**, 19 (2018)
5. D. Cho, H. J. Kim, M. J. Tahk, Journal of Guidance Control and Dynamics, **1**, 61 (2016)
6. S. Gutman Journal of Guidance, Control, and Dynamics, **8**, 2107 (2017)
7. Y. An, K. Park and C. Ryoo, *9th International Conference on Mechanical and Aerospace Engineering*, 362 (2018)
8. F. Gao, S. J. Tang, J. Shi, J. Guo, Transactions of Beijing Institute of Technology, **3**, 277 (2014)
9. H. Y. Huang, X. S. Gu, M. D. Liu, ACTA AUTOMATICA SINICA, **10**, 1115 (2007)
10. Z. J. Qi, M. D. Liu, COMPUTER TECHNOLOGY AND DEVELOPMENT, **5**, 126 (2008)